AF682625

Social Media in the Classroom

This book is part of the Peter Lang Media and Communication list.
Every volume is peer reviewed and meets
the highest quality standards for content and production.

PETER LANG
New York • Bern • Frankfurt • Berlin
Brussels • Vienna • Oxford • Warsaw

Social Media in the Classroom

Edited by Hana S. Noor Al-Deen

PETER LANG
New York • Bern • Frankfurt • Berlin
Brussels • Vienna • Oxford • Warsaw

Library of Congress Cataloging-in-Publication Data

Names: Noor Al-Deen, Hana S., editor.
Title: Social media in the classroom / edited by Hana S. Noor Al-Deen.
Description: New York : Peter Lang, 2016.
Includes bibliographical references and index.
Identifiers: LCCN 2015035392 | ISBN 978-1-4331-2905-6 (hardcover: alk. paper)
ISBN 978-1-4331-2904-9 (paperback: alk. paper) | ISBN 978-1-4539-1749-7 (e-book)
Subjects: LCSH: Internet in education. | Educational technology—
Social aspects. | Social media.
Classification: LCC LB1044.87 .S6177 2016 | DDC 371.33/44678—dc23
LC record available at http://lccn.loc.gov/2015035392

Bibliographic information published by **Die Deutsche Nationalbibliothek.**
Die Deutsche Nationalbibliothek lists this publication in the "Deutsche Nationalbibliografie"; detailed bibliographic data are available on the Internet at http://dnb.d-nb.de/.

The paper in this book meets the guidelines for permanence and durability of the Committee on Production Guidelines for Book Longevity of the Council of Library Resources.

© 2016 Peter Lang Publishing, Inc., New York
29 Broadway, 18th floor, New York, NY 10006
www.peterlang.com

All rights reserved.
Reprint or reproduction, even partially, in all forms such as microfilm, xerography, microfiche, microcard, and offset strictly prohibited.

Printed in the United States of America

To all abused, homeless, and shelter animals that are in desperate need for human compassion, support, and justice.

—Hana S. Noor Al-Deen

Table of Contents

Part Two: Applying Social Media in Teaching Public Relations

Part Three: Applying Social Media in Teaching Journalism

Assignments

Tables

Foreword

STANLEY J. BARAN

I was introduced to this timely book at a particularly interesting moment in my professional life. My Communication Department had just completed an overhaul of its writing curriculum, dropping several "old media" classes and replacing them with new courses carrying names such as Writing for Social Media, Writing for Niche Media, and Writing for Rich Media. In preparation, my colleagues and I researched the available pedagogical literature and then brought in professional writers, journalists, and marketers to help us identify what a modern writing curriculum should look like. Yes, my department's curriculum revision might have been easier to complete had we had access to these fine chapters, but what *Social Media in the Classroom* did do for us is confirm that the direction we were taking was not only up-to-date but necessary.

These chapters arrive, too, at a fortuitous time for all of us interested in the written word. Early worries that the Internet in general and social media in particular were corrupting—if not rendering obsolete—solid writing, meaningful expression, and critical thought thankfully seem to be in retreat. In their place is the idea that these technologies, like all technologies, are double-edged swords—they can be wielded for harm or for benefit. This book speaks for the benefit side of that metaphor and offers important commentary on how to best accomplish, and in fact maximize, the good that can come from understanding how to write for these new, liberating, and empowering media.

But the double edge of social media technology also cuts both ways for those of us who *teach* the use of the written word. Our students are experienced (if not necessarily skilled) media writers when they come to us already engaged in publishing their writing. This makes the classroom a particularly exciting place as we work with digital-native writers who face no barriers to publication. But this can also make the classroom a challenging place. How do we make important what has long been routine to our students? How do we generate respect for the serious work of writing for social media when it has long been personal fun for these always-on writers? How do we encourage the appreciation of the power of well-crafted words and images when passable expression has long been the minimum standard of quality necessary to meet the tastes of their 348 friends and 275 followers? How do we become better teachers of a constantly changing form of writing for a constantly changing media in an academic environment that itself is undergoing constant change? *Social Media in the Classroom* offers a fine mix of theoretical and practical approaches in answer to these questions.

For all its theoretical and practical clarity, this is also a challenging book. It challenges us to assess our assumptions about, and usual practices for, teaching writing for advertising, public relations, and journalism. It challenges us to examine and critique the philosophies and experiences of its essay authors and adjust them to meet our own pedagogical needs. Social media have changed too much about students, learning, and writing for us to maintain a safe, business-as-usual approach. This book dramatically drives this point home and expertly details a better way forward.

Preface

Social media have been growing at a tremendous rate while producing massive numbers of users worldwide. Social interaction, business interaction, and dissemination of information have been among the distinct characteristics of social media. Business organizations (such as advertising, marketing, public relations, and journalism) have consequently incorporated social media into their business operations. Such a process has led to the need for skilled staff to manage and operate social media effectively and efficiently, because social media presence has become a must for many for-profit, nonprofit, and not-for-profit business organizations regardless of the type and size. Therefore, many colleges and universities in the United States and abroad have included social media courses in their curricula as a prerequisite for advertising, marketing, public relations, journalism, and so on or as standalone social media courses while offering degrees as well as certificates in this new area of study to prepare students for the job market. Many instructors are interested in utilizing social media in their respective area of teaching, yet most often such a task might not be easily achieved due to limited resources or training.

This book can prove to be an indispensable resource for teaching the application of social media in advertising, public relations, and journalism at the undergraduate and graduate levels. It provides a truly comprehensive and tremendous source for teaching the application of social media in these areas of study by educators who were drawn from 18 prominent universities within the United States, the United Kingdom, and Australia. All of the contributors have been heavily involved in teaching the usage of social media in advertising, marketing, public relations, and journalism.

The book consists of 12 chapters of scholarly work involving teaching the application of social media in advertising, public relations, and journalism. Each chapter encompasses its own related theories, method, application, and assignment(s). Indeed, the book is comprised of 12 sections of related theories, 12 different social media teaching methods, 12 different social media applications, and 20 social media assignments/projects that were successfully applied in teaching advertising, public relations, and journalism. The assignments/projects furnish students with hands-on experience of the application of social media in these fields at the undergraduate and graduate levels. Although these three fields were selected for this volume, the teaching methods contained in this book can naturally be applied and adapted to other disciplines in the social sciences, business, education, health, and so forth. Equally important, this book can also be an excellent source for practitioners working in the fields of advertising, public relations, journalism, and so on as well as for those instructors and students who are interested in exploring such a subject area. Below are brief abstracts about the chapters in this book.

Part One: Applying Social Media in Teaching Advertising. *Chapter One: Digitally Driving Student Engagement to Improve Pedagogical Outcomes.* Marketers and educators share similar challenges in that young adults have shrinking attention spans, an increased sense of ownership and entitlement, and an increasing number of distractions such as multitasking using multiple screens or devices. Educators need to move from "pushing" content via traditional methods to utilizing the digital devices so prolifically used by students. That is, we need to use technologies relevant to our customer base in order to engage and educate. This chapter reports and reflects on the use of digital and social technologies to enhance pedagogical and theoretical educational curricula in marketing and communication content and produces relevant outcomes for graduate careers. *Chapter Two: Connectivism and the Classroom: Translating Theory into Teaching.* The foundation of this chapter is connectivism. The two-way communication capabilities of social media (in this case, blogs) provide a resource for teachers to create a networked learning environment in which students can share and make connections between new concepts. Connectivism is employed as an instructional method in a graduate-level consumer insights course for advertising, public relations, and marketing majors. A class blog is used as a tool to create a community of learning in which participants contribute information and also use information provided by others to complete an assignment. *Chapter Three: Being Your Own Chief Marketing Officer: Student Perceptions of Personal Branding.* The idea of personal branding, the practice of marketing yourself as a brand, has reached a new level of importance due to the ability of individuals to develop and promote their identities on the Internet. A strong personal brand can assist people in finding employment and obtaining promotions. A negative, weak personal brand can be disastrous for a young professional's career. Questions remain as to whether or not students recognize what a personal brand is and how it can help or hinder their careers. This chapter provides tools instructors may use to help their students in developing their personal brands and to provide information regarding student perceptions of personal branding so that curriculum can be carefully tailored to match the beliefs and expectations students have regarding personal branding. *Chapter Four: Social Media and Applied Learning.* This chapter is designed to offer students a learning approach that operates within the framework of the applied learning theory in order to empower them to transform their personal skills of social media into the advantageous use of them at a business level such as in advertising or marketing. This approach is carried out by providing students with a theoretical foundation along with practical application of using social media for advertising. The learning objective of this process is to provide students with hands-on experience in using social media to (a) promote their business and (b) maintain a business relationship with their customers.

Part Two: Applying Social Media in Teaching Public Relations. *Chapter Five: Taming the Social Media Data Deluge: Using Social Media Research Methods in the Public Relations Classroom.* Strategic communication professionals have increasingly turned to social media as a rich venue for information about key publics. The stream of comments, status updates, posts, and photos shared through popular social networking sites provides a wealth of information for public relations professionals with the research skills to make sense of the rapidly growing mass of online content. The five-week learning module described in this chapter introduces students to a variety of social media research techniques including network analysis, content analysis, influencer identification, and data mining. *Chapter Six: Drop and Give Me 20 (Social Media Platforms): Using Boot Camp to Teach Social Media Strategy.* This chapter proposes a "New Media Boot Camp" assignment to introduce public relations students to a variety of social and digital media tools. Using the theoretical frameworks of experiential learning theory (ELT) and connectivism, this chapter looks to build understanding of the importance of social and cultural contexts in implementing social media strategy in public relations and to underscore the need for models and demonstrations in effective teaching that facilitates a cyclical learning process for students. It also builds on knowledge of how to teach strategic use of social media for public relations purposes, culminating with insights and best practices. *Chapter Seven: The Art of Tweeting: Integrating Primary Social Media Research into a Public Relations Writing Course.* Research highlights the importance of two-way communication online; however, little is known about how to create dialogue with stakeholders. To address that gap, students in an upper-level undergraduate public relations advanced writing class conducted a collaborative social media research project to learn how to craft engaging messages on Twitter. This chapter illustrates how engaging in applied social media primary research emphasizes creativity and allows students to bridge the gap between theory and application. *Chapter Eight: New Technologies for Social Media and Public Relations Education.* Public relations educators of social media perform several pedagogical functions in the classroom. Not only must these educators teach new media skills, they must also be flexible enough in their teaching approach to inform both the least and most skilled students of new media. This chapter first explores the interdependent relationship between new communication technologies and social media. Next, this chapter approaches bridging the gap for issues of divide within the public relations classroom. The chapter proposes a pre- and post-assessment to identify new media skills gaps and it concludes with a discussion of the assessment and suggestion for a strategic team-based new media campaign assignment.

Part Three: Applying Social Media in Teaching Journalism. *Chapter Nine: Challenging the Newsroom Paradigm: Four Nations' Journalism Students Interrogate Global Issues Through Social Media.* This chapter maps a course beyond familiar newsroom boundaries that journalism educators might offer to their students in order to help them to explore the constantly morphing frontiers of journalism. The chapter draws on the concept of "networked journalism" to examine journalism students' responses to the introduction of a temporary, pop-up newsroom through which students from universities in four countries (India, Taiwan, the United Kingdom, and the United States) cooperatively covered the topic of poverty. Students relied primarily on cell phones and social media platforms including Twitter, YouTube, and Facebook to produce their work, often in real time. *Chapter Ten: Storytelling 2.0: Using Social Media Tools to Craft Multimedia Stories.* Journalism professors must teach students how to read critically and write effectively to prepare for careers in newsrooms. However, with the increasing usage of social media tools among professional journalists, instructors are well served to teach writing and reporting skills that use social media tools such as Twitter and Storify. Rooted within the theories of Web 2.0 and connectivism, this chapter examines the usage of a variety of social media tools (Twitter, Vine, Instagram and Storify) to enhance the multimedia storytelling experiences for journalism students and offers several examples, exercises, and student reactions from a journalism program of a southeastern regional university. The chapter is also designed to help journalists and media instructors, particularly those who teach writing and reporting skills, to build additional social media storytelling skills into their courses. *Chapter Eleven: Using Network Analytic Tools to Teach Social Media Impact on Citizen Journalism.* Citizen journalism has become increasingly prominent in the social media environment. This chapter introduces a 400-level course to exemplify how the use of the network analytic tool can be incorporated into teaching an undergraduate major on citizen journalism. The course is composed of two parts—theoretical learning on citizen journalism and hands-on training on social media network analysis. Important network analytic concepts and measurements are introduced. *Chapter Twelve: Going on a News Consumption Diet: Engaging Students in Meaningful Current Events Discussions Through Social Media.* Journalism students—perhaps some faculty—dread current events quizzes. Students must read news to write news, but today's students typically do not know what is happening in the world around them. This chapter addresses how to move away from the quiz and use social media creatively for current events assignments. Undergraduate students in a basic news writing course and an intermediate copy editing class at a regional university in the Southeast followed current news events more closely with the integration of social media techniques in the classroom. This case study applies the engagement pyramid social media model as the guiding framework to explore how social media are used pedagogically by both the instructor and the student to encourage rich discussion of current events in journalism courses.

Acknowledgments

The editor wishes to express her thanks to the contributors of this book for their enthusiasm and hard work throughout the entire process that made this book possible. Moreover, she would like to extend her sincere thanks to the anonymous reviewers as well as to the endorsers for their support of this project.

The editor wishes to express sincere gratitude to Mary Savigar, senior acquisitions editor at Peter Lang Publishing, who was extremely supportive of this project and who provided professional and unequivocal assistance throughout the entire process. Genuine appreciation is extended to Sophie Appel, design and production supervisor at Peter Lang Publishing, for her meticulous and diligent work on this project. A special thanks is conveyed to Stephen Mazur, Phyllis Korper, and Bernadette Shade at Peter Lang Publishing, for their praiseworthy efforts.

Lastly, but perhaps most importantly, the editor wishes to express her heartfelt gratitude to her husband, Omar, for his unwavering support of this project and to her four-legged Yorkie, Hercules, for his devotion in waiting for his two-legged buddy during the wee hours of the night while she was working on this volume.

—Hana S. Noor Al-Deen

PART ONE

Applying Social Media in Teaching Advertising

CHAPTER ONE

Digitally Driving Student Engagement to Improve Pedagogical Outcomes

GWYNETH HOWELL AND ROHAN MILLER

Marketing, advertising, and communication educators face unprecedented challenges as traditional lecture-tutorial methods are transitioned to incorporate or be supplanted by Web 2.0 technologies. A balance needs to be struck for student learning between interpersonal communication, engagement, and knowledge transfer using digital interfaces. Indeed, foresight and planning into the integration of digital and future learning practices should be developed in a manner superior to a university's building plans as technology is clearly going to be the dominant driver in tertiary education (Guri-Rosenblit, 2010). Arguably, digital learning technologies and back-end infrastructure are already more important in student learning than physical campus structures.

Contemporary marketing, advertising, and communication practices are a perpetual work in progress, rapidly changing with the continuous innovation and introduction of new technologies. While Internet and social media are expected to increase over time, it is difficult to foresee an era when digital (or subsequent) technologies totally supplant face-to-face interaction in most consumer and business journeys. The use of technologies by industry may provide insight into the tertiary education journey that would be presented to students: many educational models would likely incorporate some face-to-face engagement supplemented with an array of digital tools.

However, the roadmap for the future of tertiary marketing, advertising, and communication education seems less conceptually and practically developed than in the "real world" students confront on graduation. So far, "cutting-edge" learning resources such as websites that "feature" host chat rooms, teaching materials, and repositories for other miscellaneous resources do little to enhance students' learning. Yet, in many institutions these products have become an end in themselves, rather than a means to improve student education. There has been better technology available outside the university system that could serve to provide enhanced marketing, advertising, and communication learning. Moreover, why make students use software that is irrelevant after they graduate?

Indeed, just as there have been increasing numbers of consumers of all description across the globe moving between various devices and media, this same experience can be harnessed in the provision and consumption of educational content. It is paradoxical that educators seeking to produce graduates trained to target and communicate with audiences through a variety of media do not harness those media for their students' education. This is probably because of the time, effort, cost, and supporting infrastructure and personnel required to introduce and use new technology.

This chapter reports and reflects on the use of digital and social technologies to enhance pedagogical and theoretical educational curricula in marketing, advertising, and communication content and produces relevant outcomes for graduate careers. It informs and elaborates on the concepts that learning builds on prior knowledge and is enhanced by social interaction (Good & Brophy, 2007) and that "students engage in activities through which they develop skills and acquire concepts" (Cooperstein & Kocevar-Weidinger, 2004, p. 141). Moreover, it supports the concept that interactive learning models are critical to ensuring that tertiary students have effective pedagogical outcomes (Borodzicz, 2004).

LITERATURE REVIEW

In 2014, Google defined a new class of consumer they called Generation C as "a global group of consumers who thrive on creation, curation, connection and community" (The power, 2014, para. 4). This generation was not defined by age; it was defined through connection, consumption, and creation to build unique communities. A significant proportion of students attending universities and colleges have been Generation C (Gen C). This cohort is engaged online; creating videos, writing blog posts, and posting reviews, with 9 in 10 Gen C's creating online content at least once a month (The power, 2014). This engagement and activity illustrated the aptitude and understanding Gen C natives have in the digital space. Globally, this generation is considered prolific users of social and digital media.

Marketers and educators shared similar challenges in that their consumers had shrinking attention spans, an increased sense of ownership and entitlement, and an increasing number of distractions such as multitasking using multiple screens or devices (Kulesza, DeHondt, & Nezlek, 2011; Lederer, 2012). The days when educators and marketers could "push" content via traditional methods were obsolete: the sea of digital devices that confronted educators during traditional "push" lecture times was clear evidence of this change.

More than a decade ago, Miller, Howell, and Stace (2003) identified that traditional education delivery techniques were redundant in a digital technology subject, since it inhibited students from fully grasping the theory, knowledge, and skills underpinning it. Even in more general marketing, advertising, and communication subjects, the integration of digital technology and online environments provided opportunities for better and more meaningful learning outcomes for tertiary students (Ebner, Lienhardt, Rohs, & Meyer, 2010; Jordan, 2013; Kismihók et al. 2011). As social media is defined as "a group of Internet-based applications that build on the ideological and technological foundations of Web 2.0, and that allow the creation and exchange of user-generated content" (Kaplan & Haenlein, 2010, p. 61), it was important to understand how these sites enhanced learning outcomes. The collaborative, communal, and conversational qualities of social networking sites "mirror what we know to be good models of learning, in that they are collaborative and encourage active participatory role for users" (Maloney, 2007, p. 26). This supported the Laird and Kuh (2005) suggestion that "there appears to be a strong positive relationship between using technology for educational purposes and involvement in effective educational practices such as active and collaborative learning and student-faculty interaction" (p. 211). Learning builds on prior knowledge and is enhanced by social interaction (Good & Brophy, 2007) and interactive learning models were critical to ensuring that tertiary students had effective pedagogical outcomes (Borodzicz, 2004). Students were more likely to engage in activities through which they develop skills and acquire concepts (Cooperstein & Kocevar-Weidinger, 2004).

The general consensus between marketing, advertising, and communication academics and practitioners was that interactive learning models were critical to effective pedagogical outcomes (Borodzicz, 2004). For Gen C students, the majority adopted technological skills early on in life and were well prepared for such advances in technology-based education models (Guy, 2012; McHaney, 2011). Ulbrich, Jahnke, and Mårtensson (2010) argued that "members of the net generation use the web differently, they network differently, and they learn differently" (p. ii). It was tautological that tertiary educators leveraged this understanding in the provision of educational content. Educators should behave like marketers in the adoption and use of digital channels to achieve an appropriate share of student screen-time and engagement.

Social media have been fundamental to the interactions of Gen C, X, Y, and Z (Browning, Gerlich & Westermann, 2011; The power, 2014), and the potential benefits of using digital technology for academic purposes are widely recognized

(Chen & Bryer, 2012; Hughes, 2009). However, a gap remains in the use of social media in education; there seemed to have been limited adoption of digital and social media (outside the standard "learning platforms") for educational or learning purposes (Selwyn, 2012).

Often the lack of digital engagement between students and academe was due to academics not incorporating the technology into their content delivery. In 2013, only 41% of Australian academics reported use of social media channels for teaching purposes (Tinti-Kane, 2013) yet an estimated 97% of Australian academics use social media as part of their academic work (Lupton, 2014).

There seemed to be several reasons for the social and digital media divide between consumer use and teaching practices. Chen and Bryer (2012) identified that many academics used social media personally, but time constraints, workloads, privacy issues, and cyberbullying were cited as reasons for not using these channels in academic engagement and content delivery. To better understand social media, the range of platforms types can be based on the utility they provided. The opportunity cost of not leveraging contemporary technology such as these platforms in marketing education was high. By not utilizing contemporary social and digital media in teaching, were educators adequately preparing students for their professional careers?

There was clearly an element of resistance to technological change in education models. Educators could teach the history of marketing and communications on the Internet easily using available learning materials, but what use were graduates who did not understand the current digital marketing and communication practices and the performance metrics on which success could be evaluated?

Web 2.0 added an ever-growing number and type of marketing touch points and an array of new data to marketing practice, and this created a knowledge gap about how to manage and measure the success of digital marketing and communication (Forrester Consulting, 2012). Marketing, advertising, and communication practitioners faced the challenge of keeping abreast of new and evolving performance management concepts such as attribution modelling (that in many instances supplanted the "last click" assumption), as well as the expectation in their ability to explain how and why particular metrics were used of marketing or communication success.

Setting out to impart a level of current knowledge relevant to industry has had drawbacks and problems that educators needed to navigate. A particular equity concern was the disparity between students in terms of access to, experience of, and skills in using technology. A recent U.S. study identified age, ethnicity, and gender imbalances as key factors underpinning a "digital divide" between users and non-users of technology that has rapidly increased (BrckaLorenz, Haeger, Nailos, & Rabourn, 2013). While technology can be utilized to assist students learning in languages other than their native tongue (Miller et al., 2003), the globalization of the education industry suggested that foreign students can be at a major disadvantage when localized digital technology is used for teaching. Thus while there are possibilities for broadening the opportunities for teaching and learning through innovative techniques, educators should also consider the potential challenges in crossing the digital divide.

One challenge facing educators has been that digital technology could change in a matter of weeks, and without notice. Thus, learning or assessment developed at any time prior to the class might be obsolete and unworkable during the semester.

From the educational consumers' perspective, the active use and integration of digital technologies into courses meant students were required to shift from their comfortable paradigm as a user of selected technology and asked to take a different outlook (for example, view the world as a marketer rather than a consumer). Moreover, students may be asked to learn and use the same technologies that have been applied in industry; however, these are characterized by developmental learning materials and ever-changing technology. Students may also be taught new theory, such as the law of increasing returns (Arthur, 1996), which challenged traditional theory. Thus, educators using applied or embedded learning technology are likely to move many of their students from their traditional comfort zone in which knowledge was enshrined in textbooks and selected readings to an ever-changing hands-on learning paradigm.

Shifting the Teaching Paradigm to Co-creation

The difficulties of integrating digital technologies into the curricula could be offset by the benefits of the technology, and in particular, the gain achieved by "relationship-building between faculty and students participating in this learning curve together and increased student interest and creativity" (Morrone, Gosney, & Engel, 2012, p. 6). However, the dynamics of digital technology implied educators were co-creators and actively learning along with their students. And yet, educators might not necessarily know more than their students in all aspects of digital technology, particularly in advanced classes where students have had firsthand knowledge using digital technology in industry. The shift to co-creator was likely a novel and (perhaps) daunting proposition for educators and students alike. However, the reality was that in the ever-changing digital world, it was impossible to keep abreast of all possibly relevant digital marketing and communication materials.

Educators spending time as visiting professors and taking sabbaticals with industry are not new concepts, but they would likely become more common in the future as this is one way that educators could learn and keep abreast of technology. Similarly, MOOCs created by industry and targeted to educators would likely become more prevalent; HootSuite has been an obvious example of an organization seeking to partner with educators and provide educators and students with materials to benefit their respective roles in the co-creation of knowledge in the digital environment. Another way to develop knowledge is working with industry in research partnerships to better understand when and how to use digital and social media (e.g., Harvest Strategy's 2014 Social Games work).

This chapter supports the steadfast principle that teaching practices should be driven by educational objectives rather than technological desires (Freeman & Capper, 1999; Kirkup & Kirkwood, 2005) and suggests this means cutting-edge technology must be proactively adopted and used, especially as teaching has become an activity in co-creation. If the conceptual and practical basis of education content could be better understood using an applied multimedia approach, then digital media need to be integrated with the core element of subject design (Kimmons & Veletsianos, 2014). In other words, if marketing, advertising, and communication educators want to practice what they preach, a selection of the digital platforms used by industry should be used to transition students into engaged "prosumers" and co-creators of theoretical and applied content. Digitally based learning tools in the form of specifically designed online tasks such as a digital communication strategy for a real-world client could shift students to more actively participating in the construction of knowledge and learning (Bonk & Kim, 2006; Sweeney & Ingram, 2001). Appropriate online learning could enhance student learning, better preparing students to make effective use of technology in their post-education workplace and enabling educators to be more productive (Freeman & Capper, 1999; Miller et al., 2003). As interactivity was a teaching tool commonly thought to improve learning outcomes (Guy, 2012; Jonassen, 2000), there was strong pedagogy to apply digital and social media technology to enable students to use interactive technology-based experiential learning. This could be achieved by moving students beyond "techno-competency" skills (Smart, Tomkovick, Jones, & Menon, 1999) and into a world of digital interaction that enabled educational content and theory to be learned and practiced. The use of social media could provide a superior and engaged learning experience for tertiary students, while also considering the various learning styles required by students (Lupton, 2014).

METHOD

This chapter seeks to provide educators with insights to enhance the successful implementation of social and digital media within marketing, advertising, and communication education. It does this by providing a deeper understanding of emergent themes and practices identified in the research findings. Based on insights from qualitative and quantitative data from two Australian universities, this chapter provides a snapshot of some of the issues faced by educators when implementing and operating social and digital technologies within university degree programs.

The central proposition of this chapter is that learning practices for digital marketing, advertising, and communication education need to be customized for digital natives using experiential and embedded learning techniques. The outcomes from this research illustrated that fundamental readjustment was required in our educational business models to integrate digital technology so as to provide students with the theory and practical knowledge they would need to be able to evaluate critically and use technology in the professional careers.

The content for the units of study was delivered via a range of media including face-to-face lectures and workshops, lab sessions, online lectures, collaborative sessions (real-time online software), and social media discussions. A class YouTube channel was also created to present, develop, and manage content throughout the semester. Class feedback suggested the YouTube channel produced strong learning outcomes and motivated the cohort's interest as it brought relevancy to theory and case studies in a contemporary environment. This was a logical extension to Fill and Ottewill's (2006) earlier suggestion that "video clips can provide a compelling and immersive educational experience and are able to stimulate the students' interest" (p. 397). Further, employing the Cooperstein and Kocevar-Weidinger (2004) suggestion that "a review of the key concepts immediately at the end of the session is the single most important factor to fix information in the long-term memory and cut the forgetting curve" (p. 145), at the end of each lecture, a "vox-pop" clip was produced by the educator and uploaded into Facebook to remind the students of the key aspects of the lecture and materials covered.

APPLICATIONS

To learn more about Australian marketing students' digital usage and to provide students an opportunity to reflect on the potential impact digital media may have on marketing, advertising, and consumption, students were asked to keep a digital diary as part of their assessment in digital marketing classes in two universities. In general, the students' use of digital media was omnipresent across time, place, and devices, thereby making it extremely difficult for many students to keep accurate diaries.

Australian Tertiary Student Online Engagement

Each "learning" market has its own characteristics for digital use and engagement. For example, on average, 99% of Australians aged between 20 and 29 years use Facebook 33 times per week, for 18 minutes per visit, to contact their 366 Facebook friends, but only ever see 41% of their "friends" during a year (Yellow, 2013). This was a similar age demographic found in Australian tertiary institutions, which might imply that many Australian tertiary students were somehow found in the vicinity of 594 minutes (or almost 10 hours) each week to engage on Facebook. It was worth noting that time spent on Twitter, Google+, Instagram, and the many other digital products was not considered in these usage data. Further, the 20- to 29-year-old demographic was extremely mobile when accessing digital content, with 86% reporting the use of smartphones, 69% using laptop computers, and 29% using tablets compared with 36% using desktop computers (Yellow, 2013).

Australian Tertiary Student Digital Native Profile

The data used in this study were drawn from undergraduate and postgraduate populations undertaking units of study that required students to complete a personal digital diary. The students' average total weekly digital media use was 60.06 hours per week ($N = 320$). The time allocated to "personal" Facebook usage accounted for 10.06 hours, which was similar to the 2013 Yellow social media report. In addition, 4.4 hours each week were spent "texting." Over 25% of the students' overall time on digital media (15.67 hours per week) was spent on university-related readings and postings, real-time discussions, and other forms of engagement with fellow students. While the data were self-reported, they illustrated how students were participating in digital media, and the time allocation in terms of student engagement and learning.

Communication. A fundamental part of an educator's job has been to facilitate learning, and this was typically achieved by face-to-face communication. The introduction of web-based technology, such as Blackboard, provided a forum through which educators and students could "formally" connect. However, Blackboard was not useful once students graduated, and it did not seem to fit within an immersive learning model, other than to provide a tool for "formal" communication.

Thus, to supplement face-to-face and Blackboard communication, educators could use less formal communication such as a combination of Facebook, Twitter, WordPress blogs, Linkedin, Instagram, Tumblr, and YouTube, typically supported with HootSuite. Research libraries and document repositories outside the university infrastructure can be created in Google Docs, Dropbox, and DIIGO, and these can be augmented with other resources such as Google AdWords, Google Analytics, SEMrush, SocialMentions, Moz, and Social Bakers. These real-world tools are employed to increase engagement, enhance learning outcomes, improve peer-to-peer discussion, and provide informal student discourse. Additional information, such as links to topical and useful publications, YouTube content, Slideshare decks, blogs, and website links, are uploaded by all in the class.

Veletsianos and Kimmons (2013) found that many educators did not want to position themselves as their students' friends or equals, or to engage in personal discussions, a key aspect that underpinned Facebook's success. To manage this situation, academic avatars were employed, and these produced impressive engagement and outstanding pedagogical results. Each educator's avatar was derived from the title of the unit taught, and all students were aware that the avatar was managed by the staff member concerned. Students could use their own personal Facebook account to "friend" the avatar or elect to develop their own educational avatar for the duration of the unit. The majority of students used their own personal Facebook account, happy with the academic's undertaking that personal information posted on the individual's Facebook feed would not be commented on by the academics. Students were encouraged to post into closed Facebook groups established by the academics teaching into the units. Each semester the avatars enjoyed 100% engagement as all the students in each class joined the semester's private group on Facebook. Some students, however, opined that using their Facebook page to engage in academic forums made their feed cluttered, and that regardless of privacy settings, some students also felt their academic life was crossing a precious boundary into their personal life. Facebook clutter could be particularly relevant if multiple subjects were utilizing the same digital communication channels. The use of unit hashtags assisted in the filtering process and HootSuite enabled better feeds containing only relevant information.

On average, 75% of the students in these units followed the educators on Twitter and directly tweeted about the course, reflecting on their actions and responses as well as the course materials and topics covered in the weekly lectures. The use of Twitter in communication between educators and students has evolved since 2012, and engagement in 2014 demonstrated tweets were a vibrant multidirectional channel. In particular, Twitter has been useful for questions posed in real time, note-making of key aspects of the teaching materials, the sharing of resources, and the establishment of an online presence for the students and their colleagues.

Social media provided a forum for educators to chat online about issues raised in class, posted status reports on current issues (for example, from the media or digital information stream), as well as the topics covered in the readings. Further, social media provided another way for educators to engage in discussions about complex concepts explored in the readings and to post other useful information so students gain further insight into their projects, theoretical frameworks, and communication. "For the most part, [Facebook] was AWESOME because I had a great place to discuss things with people and find information & updates. A BILLION times better than Blackboard or email notifications!" (M. Vella, personal communication, 2014). Digital and social media could be particularly useful for international students who might not speak English as their first language and needed time to ruminate or obtain the real meaning of key words and concepts. At times, students forgot they were chatting to an educator; they asked questions and probed topics and enjoyed the social media platform far more than in the formal university teaching site. As students employed a mix of social media tools, they were required to hashtag all their engagement with the unit code and date as this formed a method to illustrate and track engagement and participation in the unit. Educators acted as moderators for posts and provided support to groups and individuals struggling with topics in real time, thus enhancing their understanding and learning.

Assessments. The assessment components used include an individual digital diary (loaded daily into a spreadsheet, followed by a related reflection task), a range of tutorial tasks linked to online platforms such as HootSuite Certification and Google Analytics, a digital marketing and communication strategy developed as a blog, and an executive summary produced as a YouTube clip (see Assignment 1.1). Consistent with the principle that meaningful learning was developed through "authentic" tasks (Good & Brophy, 2007), as part of the digital strategy, students were briefed by a real-world client who required a digital marketing and communication strategy. Students responded well and recognized the value of the assessment and this approach. Further, students acknowledged the increased workload for the unit: "it was hard work but we thoroughly enjoyed it! Will take with us lots of useful skills to apply in our future careers too. Thank you for putting together and executing such an interesting course" (F. Tran, personal communication, 2014). The briefings typically occurred via Skype or Hangout sessions, site visits, and formal discussions.

On completion of the project, the students returned to the classroom to analyze their progress and present their strategies to the class and the "client." With guidance from the educator, the client judged the best campaign and awarded a team prize to the winners. Students reflected on the project: "it was so interesting to see how your group could make something with just an idea. The digital platforms were a pain, but I'm glad we were taught them. Not only does it look good on a CV but it also makes me more confident when applying for positions" (S. Mitrovic, personal communication, 2014).

So, rather than studying "how" to develop a digital and social communication strategy in a static environment, students were required to rethink the polarized dynamics of traditional marketing, advertising, and communication strategy and content, and develop "live" plans rather than following prescriptive checklists. "I really enjoyed being able to develop our own campaign. That 'hands on' approach really makes you feel like your accomplishing something and taking a step towards the workforce—something I wish more units could do!" (M. Vella, personal communication, 2014). Further, the unit empowered students to gain experience that employers have been seeking: "I was successful in getting the internship for an SEO company, I use Google Analytics daily so it was great to be able to say I had a brief understanding of it. I also now use Hootsuite to manage all of their social media pages, so I must say that it has been by far the most industry relevant unit I've done to date (and one of the only things I've disliked about unit is how irrelevant some of the units seem)" (R. Tomlinson, personal communication, 2014). The impact of student engagement throughout the project by building content, blogging the strategy, and producing the "executive summary" as a two-minute YouTube clip ensured they continually returned to real-world outcomes that align with the principles of critical theory for teaching and learning (Morrow & Torres, 2002).

While the students could be assessed on their individual postings and engagement, as well as their work as a team, they were also required to reflect on their actions and responses. However, digital participation could be difficult to assess and consideration needed to be given to the quality of posts and discussion, compared to a mere "like" or banal comment. The initial intake observed, "the fact that people were spamming it [closed Facebook page] with stuff made me get annoyed with the page, and as a result as time went by I started to ignore it more and more, which led to me missing some very important posts. I'd suggest that monitoring our online activity be shown another way, like a report or something, instead of the use of that page" (M. Vella, personal communication, 2014). These differences could be difficult to enunciate to students for whom the shift into assessable digital communication was new. The sheer volume of posts could also make assessment difficult: tools such as Hootsuite could be valuable in these circumstances.

Hence, the third time the course was delivered, students were required to undertake a digital treasure hunt. Rather than students being assessed for their social media engagement throughout the entire semester, students were assessed over a seven-day block. During this time, they were required to undertake a digital treasure hunt that enabled students to illustrate their understanding and ability to use the range of technological tools that they were supposed to have worked with during the semester. In the treasure hunt, students were required to find six clues and complete the tasks associated with each clue, followed by a concise 300-word reflection of the experience. This short time frame drastically reduced the "long-term clutter" in the student's social media feed—a general concern held by students. Further, the week-long task enabled students to demonstrate their abilities and to experience high levels of marketing and advertising activity in a closed group. The treasure hunt was widely embraced and accepted by students, produced excellent learning outcomes, and improved assessment/grading consistency: "The digital treasure hunt showed me how much better advertisers must become if they use Facebook to reach me as a customer." Ultimately, the future of digital participation marking would lie in the adoption of technology that would improve and automate marking rubrics.

DISCUSSION

For students, integrating a suite of digital and real-world tools into learning content can be a major departure from the usual teaching and learning materials, and does come with risks. However, when done well, many students recognized the benefit of the innovative curriculum drawing from real-world digital tools: "I regard this unit as being the most valuable unit I have taken as part of my degree" (Anonymous, personal communication, 2013). In addition to the real-world digital strategy project, the curricula were designed to provide a cutting-edge learning environment that was assured by rigorous assessment. Each week tweets, status updates, online event invitations, the week's lecture materials (explicitly including theory), and discussions of current digital "stories" provided scope for student learning. The courses received comments such as, "This course taught me a lot, and it's actually been the most important one to draw from at my internship. In terms of learning content, it's doing pretty well" (C. Collins, personal communication, 2014). Students typically embraced the range of communication and engagement tools—"love your tweets, gets me excited about class!" (Anonymous, personal communication, 2013).

The evidence from students suggested that they welcomed the challenge of the online strategy and the aligned learning projects and these were highly effective teaching tools. "I really enjoyed being able to develop our own campaign. That 'hands on' approach really makes you feel like your accomplishing something and taking a step towards the workforce—something I wish more units could do!" (M. Vella, personal communication, 2014). Social and digital tools with Blackboard support in the background shifted students to become more active participants in the construction of knowledge and learning. The students appreciated that the project's digital outputs could be used as "a portfolio of work, its something i could show and tell rather than just tell" (Z. Lord, personal communication, 2014). Further, "I enjoyed the subject, but I have to say thanks to you, your teaching style was engaging and motivating" (R. Blackstone, personal communication, 2014).

Social media can and does provide the opportunity for academics to develop interactive and engaging content and assessment for students and to assist in the creation of learning communities. Digital media such as Facebook, Twitter, and Tumblr allow educators and our consumers to share and comment on information, data, and lecture materials, interact in real time with both peers and educators, and generate learning materials. The engaging and interactive nature enabled by digital media creates an opportunity for academics to convert students into prosumers of knowledge throughout the course of the semester, and to help them develop a peer network within their area of study.

Digital technology is here and it will only become more pervasive. The onus is on educators and students to embrace innovation and learning opportunities. However, it must be recognized that digital technology continually evolves and it requires considerable time and effort to stay abreast of these fast-moving trends. University administrators must provide more resources and redefine workload models to encourage and enable educators to harness digital technology for constructed pedagogical outcomes that prepare students for the real-world challenges they will face.

REFERENCES

Arthur, W. B. (1996, July–August). Increasing returns and the new world of business. *Harvard Business Review*. Retrieved from https://hbr.org/1996/07/increasing-returns-and-the-new-world-of-business

Bonk, C. J., & Kim, K. (2006). The future of online teaching and learning in higher education: The survey says. *Educause quarterly—educause.edu. 4*. Retrieved from http://www.educause.edu/ero/article/future-online-teaching-and-learning-higher-education-survey-says%E2%80%A6

Borodzicz, E. P. (2004). The missing ingredient is the value of flexibility. *Simulation and Gaming, 35*, 414–426.

BrckaLorenz, A., Haeger, H., Nailos, J., & Rabourn, K. (2013). *Student perspectives on the importance and use of technology in learning*. Paper presented at the Annual Forum of the Association for Institutional Research, Long Beach, California. Retrieved from http://cpr.iub.edu/uploads/NSSE13%20AIR%20Technology%20Paper.pdf

Browning, L., Gerlich, R., & Westermann, L. (2011). The new HD classroom: A hyper diverse approach to engaging with students. *Journal of Instructional Pedagogies, 1*(10).

Chen, B., & Bryer, T. (2012). Investigating instructional strategies for using social media in formal and informal learning. *The International Review of Open and Distance Learning, 13*(1). Retrieved from http://www.irrodl.org/index.php/irrodl/article/view/1027/2073

Cooperstein, S. E., & Kocevar-Weidinger, E. (2004). Beyond active learning: A constructivist approach to learning. *Reference Services Review, 32*(2), 141–148.

Ebner, M., Lienhardt, C., Rohs, M., & Meyer, I. (2010). Microblogs in higher education—a chance to facilitate informal and process-oriented learning. *Computers & Education, 55*, 92–100.

Fill K., & Ottewill, R. (2006). Sink or swim: Taking advantage of developments in video streaming. *Innovations in Education and Teaching International, 43*(4), 397–408.

Forrester Consulting. (2012, June). Digital attribution comes of age: The Last-Click paradigm erodes as marketers turn to fractional attribution, commissioned. *The Interactive Advertising Bureau*. Retrieved from http://www.iab.net/media/file/DigitalAttributionComesOfAge.pdf

Freeman, M. A., & Capper, J.A. (1999). Exploiting the web for education: An anonymous asynchronous role simulation. *Australian Journal of Educational Technology, 15*(1), 95–116.

Good, T., & Brophy, J. (2007). *Looking in classrooms* (10th ed.). New York, NY: HarperCollins.

Guri-Rosenblit, S. (2010). *Digital technologies in higher education: Sweeping expectations and actual effects*. Israel: Nova.

Guy, R. (2012). The use of social media for academic practice: A review of literature. *Kentucky Journal of Higher Education Policy and Practice, 1*(2). Retrieved from http://uknowledge.uky.edu/kjhepp/vol1/iss2/7

Harvest Strategy. (2014, May). A new industry profile: Digital+Social+Gaming=Disgsogames. *Research Report*. Retrieved from http://www.i-sga.org/research/

Hughes, G. (2009). Social software: New opportunities for challenging social inequalities in learning? *Learning, Media and Technology, 34*(4), 291–305.

Jonassen, D. H. (2000). Toward a design theory of problem solving. *Educational Technology Research and Development, 48*, 63–85.

Jordan, L. (2013). *A comparative analysis of student engagement between community college students* (Doctoral Dissertations and Projects). Retrieved from http://digitalcommons.liberty.edu/doctoral/722

Kaplan, A. M., & Haenlein, M. (2010). Users of the world, unite! The challenges and opportunities of social media. *Business Horizons, 53*(1), 59–68.

Kimmons, R., & Veletsianos, G. (2014). The fragmented educator 2.0: Social networking sites. Acceptable identity fragments, and the identity constellation. *Computers & Education, 72,* 292–301. Retrieved from http://www.veletsianos.com/publications/#sthash.gHAmS22C.dpuf

Kirkup, G., & Kirkwood, A. (2005). Information and communications technologies (ICT) in higher education teaching—a tale of gradualism rather than revolution, learning. *Media and Technology, 30*(2), 185–199.

Kismihók, G., Mol, S., Vas, R., Duijts, P., Costello, V., Sorrentino, G., & … Szabó, I. (2011). An innovative e-learning system for competence matching between vocational education and the workplace. *Proceedings of SKIMA.* Presented at the SKIMA, Benevento, Italy.

Kulesza, J., DeHondt II, G., & Nezlek, G. (2011). More technology, less learning. *Information Systems Education Journal, 9*(7). Retrieved from http://isedj.org/2011-9/N7/ISEDJv9n7p4.pdf

Laird, N., & Kuh, G. (2005). Student experiences with information technology and their relationship to other aspects of student engagement. *Research in Higher Education, 46.*

Lederer, K. (2012). Pros and cons of social media in the classroom. *Campus Technology, 25*(5), 1–2.

Lupton, D. (2014, June 10). *"Feeling better connected": Academics' use of social media.* Retrieved from University of Canberra, News & Media Research Centre, http://www.canberra.edu.au/about-uc/faculties/arts-design/attachments2/pdf/n-and-mrc/Feeling-Better-Connected-report-final.pdf

Maloney, E. (2007). What Web 2.0 can teach us about learning. *Chronicle of Higher Education 53*(18), B26.

McHaney, R. (2011). *The new digital shoreline: How Web 2.0 and Millennials are revolutionizing higher education.* Sterling, VA: Stylus.

Miller, R., Howell, G. V. J., & Stace, R. (2003). A developmental approach to teaching Internet marketing. Australian and New Zealand Marketing Academy Conference, Adelaide SA, Australia.

Morrone, A. S., Gosney, J., & Engel, S. (2012). Empowering students and instructors: Reflections on the effectiveness of iPads for teaching and learning. *EDUCAUSE: Learning Initiative.* Retrieved from https://net.educause.edu/ir/library/pdf/ELIB1201.pdf

Morrow, R. A., & Torres, C. A. (2002). *Reading Freire and Habermas: Critical pedagogy and transformative social change.* New York, NY: Teachers College Press.

Selwyn N. (2012). Social media in higher education. *The Europa World of Learning.* Retrieved from http://www.educationarena.com/pdf/sample/sample-essay-selwyn.pdf

Smart, D. T., Tomkovick, C., Jones, E., & Menon, A. (1999). Undergraduate marketing education in the 21st century: Views from three institutions. *Marketing Education Review, 9,* 1–9.

Sweeney, J. C., & Ingham, D. (2001). A comparison of traditional and web-based tutorials in marketing education: An exploratory study. *Journal of Marketing Education, 23*(1), 55–62.

The power of Gen C: Connecting with your best customers. (2014). Retrieved from http://www.thinkwithgoogle.com/articles/the-power-of-gen-c-connecting-with-your-best-customers.html

Tinti-Kane, H. (2013, April 1). Overcoming hurdles to social media in education. *educausereview online.* Retrieved from http://www.educause.edu/ero/article/overcoming-hurdles-social-media-education

Ulbrich, F., Jahnke, I., & Mårtensson, P. (2010). Special issue on knowledge development and the net generation. *International Journal of Sociotechnology and Knowledge Development,* 2(4), i–ii.

Veletsianos, G., & Kimmons, R. (2013). Scholars and faculty members lived experiences in online social networks. *The Internet and Higher Education, 16*(1), 43–50. Retrieved from http://www.veletsianos.com/publications/#sthash.gHAmS22C.dpuf

Yellow social media report. (2013). Retrieved from http://about.sensis.com.au/IgnitionSuite/uploads/docs/Yellow%20Pages%20Social%20Media%20Report_F.PDF

ASSIGNMENT 1.1. DIGITAL STRATEGY PROJECT

Aims/Objectives

To develop an innovative Digital Strategy to address the issues raised in the initial briefing notes and subsequent information collated. The digital approach is to encompass all aspects of the development and implementation of the strategy, creative, media planning, communication channels, content, as well as analytics to measure impact, demonstrating an understanding of the integration of the various digital and social media platforms available.

Assessment Description

In groups of four or five you will develop an innovative Digital and Social Media Strategy to address the issues raised by the client in the initial briefing. The entire strategy is to be developed for an online format with the overall budget of $10,000. Your group needs to allocate specific roles to each member of the group, then work to integrate individual activities into the final plan.

Assessment Requirements

The overall strategy executive summary video pitch to be submitted on Blackboard will include the creative, media, online, and communications elements of the program. This will be supported by the digital blog: the blog will cover the areas of research, stakeholders, strategy, creative/tactics, rational, outcomes, and digital and social media channels and platforms. Further links to all aspects of the project will be provided in the summary.

Assessment Criteria

A 2,500-word equivalent online strategy blog, and a Video Executive Summary/Pitch. See marking rubrics on Blackboard for greater detail.

Executive Summary 10%
Two minutes TOTAL Video Presentation

You can choose your own manner for the production of the video. The video will be the "sales pitch" to the client as to why your campaign is the best one to choose. All videos will be uploaded onto YouTube and all videos will be compiled as a show reel for the client.

Group Blog: Strategy Development, Full Campaign 20%
Created in WordPress and links shared with academics in Blackboard

1. Undertake an introduction comment into the group blog that *must* be secured with *password* access to group and academics. Each group will establish and engage using the blog to track all group interactions and planning during the development of the project. Illustrate strong digital and social engagement with each other and drawing on content and information shared by the academics and guest lecturers during the semester.

Remember:

- EVERY TEAM MUST PROVIDE PASSWORD TO TUTOR IN WEEK 4
- ANY TEAM MEMBER FOUND TO BE SHARING THE GROUP'S INFORMATION WITH OTHER STUDENTS WILL RECEIVE 0/50 FOR THE ASSESSMENT.

2. Each group will complete the Business Objectives Plan as per Google Analytics and post into blog.
3. You need to develop the content for the various platforms to be used in your campaign. This includes an editorial calendar for the next six months.
4. The campaign is to incorporate the various digital platforms that best reached their target. As CLIENT is an existing company, you need to create the various platforms, load them up with content, and not send them live, please make your tutor has admin for the site so that we have access.
5. Consider the platforms needed to drive the campaign while working together to deliver the campaign strategy that was based on the research (target audience, segmentation, competition, etc.). You need to develop clever campaigns that included the digital and physical worlds.
6. You are to create and execute within the limitations of closed site, then pitch your idea and blog about the communications plan (covering research, strategy, creative and digital platforms) with relevant links. If part of your plan is a website or landing pages for competitions, etc., you need to go as far as creating the wireframes.
7. Measurement of the campaign needs to be clearly articulated, drawing on the skills developed during the unit.
8. Submit the program/strategy/creative/content/measurement.

Marking Criteria

See Blackboard for marking criteria.

CHAPTER TWO

Connectivism and the Classroom

Translating Theory into Teaching

REGINA LEWIS AND BRANDI WATKINS

The emergence of Web 2.0 technology has left an indelible mark on society, affecting even the ways in which instructors might choose to redesign curricula. It has been well-documented that graduates entering fields related to advertising, public relations, and all avenues of marketing must have a full comprehension not only of traditional communication channels but also of emerging platforms such as social media and mobile applications.

While the majority of changes to communication and marketing curricula have included new, stand-alone courses specific to digital media design, website monitoring, web content development, and other digital topics, many instructors teaching traditional courses have recognized that new media must be seamlessly incorporated into their currently existing classes, as well, if students' educations are to be holistic enough to serve them in the evolving workplace. Put another way, to meet new and upcoming technological demands of the professions, instructors must incorporate new teaching strategies across the board to further familiarize students with new media. A change in workplace practices necessitated change in instruction, which in turn requires a change in the approach to learning. And this, in turn, poses many challenges to developers of curricula such as those associated with historically generalist advertising programs. Inserting new tools into classrooms can be easy; however, inserting the right tools in the right way—such that student learning truly is enhanced—can be quite difficult.

Among other new media additions to typical communication and marketing syllabi, which have included the basics of search engine marketing and best practices with regard to Twitter, Facebook, and Instagram, have been best practices with regard to blogs. Blogs, which typically exist in the form of websites that express one or more writers' input or observations on a topic, represent their own category of media tool. The incorporation of blogs into communication and marketing courses typically has involved a requirement that individual students envision and write blog content as a way of demonstrating their ability to build credibility through social media. Or it has involved a requirement that students develop blogs to demonstrate copywriting or creative skills. The exercise behind the focus of this chapter, however, involved the sharing of collective small-group knowledge bases via blogging in the classroom for the sole purpose of improving overall class knowledge and expertise about a topic. By focusing on blogging in this fashion, this chapter contributes to writings on the theory of connectivism.

Connectivism has provided a framework for adjusting pedagogy in a way that reflected the influence of new technology on the communications and marketing industries. So, the foundation of this chapter is connectivism, a theory positing that learning is a process of connecting specialized information sets to further knowledge (Siemens, 2005). As data were gathered to examine and illustrate potential information points within this chapter, connectivism was employed as an instructional method in a graduate-level consumer insights course for advertising, public relations, and marketing majors. The class also included some students focusing on other areas of communication, such as telecommunication and film, but these students represented the minority.

The purpose of this chapter is to understand whether, in a consumer insights and understanding course, new client discovery can be made more robust by asking students to create information sets, build upon them, and then make them available via social media to classmates serving in mock agency roles. Because the availability of student-created blogs appeared to have heightened productivity in information gathering for hypothetical new clients, it would seem the author(s)' exploration was worthwhile. And, it would seem that many other ways to both engage students with social media and heighten learning in the classroom might be possible.

LITERATURE REVIEW

Connectivism

Siemens (2005) recognized that traditional learning theories, such as behaviorism, cognitivism, and constructivism, did not reflect the impact of technology in the classroom. In a 2006 article, he argued,

> we communicate differently than we did even ten years ago. We use different tools for learning; we experience knowledge in different formats and a different pace. We are exposed to an

> overwhelming amount of information—requiring continually greater levels of specializations in our organizations. (pp. 9–10)

As technology has provided a way to generate and access new information and data at a staggering rate, there should be a way to organize and process this information. Additionally, Siemens (2005) argued that making connections from the various data sets is an integral piece of the learning process. Therefore, the theory of connectivism is a knowledge management theory, which aids in managing new sources of information (Siemens, 2005). This has been especially relevant in strategic communication fields such as advertising, marketing, and public relations, which have been increasingly becoming data-driven fields that relied on research and consumer insights to develop communication campaigns.

Siemens suggested that the learning environment should be reflective of the learner's social environment; thus, the new technology-driven social environment required new ways of thinking about learning. New trends in learning suggest that "(1) learning is a continual process, (2) informal learning plays a large role in the learning experience, and (3) technology is 'altering our brains' and affecting the way people process information" (Siemens, 2005, p. 1). Learning and generating new knowledge can be viewed as a social activity in which people would take in and contribute new ideas (Bell, 2010). New technology that provided opportunities for two-way communication created a new space for learning that allowed for a free flow of information. However, as technology allowed for an increase in the amount of knowledge and information generated, as previously mentioned, there existed a need to organize and manage that information. In light of this, Siemens (2005) suggested that educators have emphasized making connections to enhance the learning experience.

Connectivism, as a learning theory, emphasized making connections between existing and new information (Siemens, 2005). "Learning is a network phenomenon, influenced (aided) by socialization and technology" (Siemens, 2006, p. 4). In the connectivism approach, knowledge is distributed across a network and through learning the process of seeing patterns emerge from those networks (Siemens, 2006). Kop and Fournier (2010) added, "being a member of an online network, communicating with others, and filtering information and ideas that others provide will lead to knowledge creation and learning advancement" (p. 4). The basic idea is that decisions are based on rapidly changing knowledge foundations and new information is constantly being acquired. Therefore, being an active member of a network and making connections between new and existing information has been an integral part of the learning process.

Blogs in the Classroom

Blogs have been a staple of the Web 2.0 technologies used in the classroom (Higdon & Topaz, 2009). Lapadat, Brown, Thielmann, and McGregor (2010) defined blogs as "individual or group websites that can be easily created, customized, updated and edited" (p. 64). Blog posts are archived in reverse chronological order and contain text, hyperlinks, and multimedia content (Lapadat et al., 2010). Mak, Williams, and Mackness (2010) suggested that blogs offered students a social presence and a channel for connection, self-expression, and self-indulgence, among other activities. Often, the "permanence" of information available on blogs made them a valuable resource for aggregating class-related information (Martindale & Wiley, 2004).

Duffy and Bruns (2006) listed several educational benefits of blogs, including "(a) promote critical and analytical thinking, (b) promote creative, intuitive and associational thinking, (c) promote analogical thinking, (d) potential for increased access and exposure to quality information, and (e) combination of solitary and social interaction" (p. 33). Churchill (2009) suggested that blogs can be used to publish student writing, coordinate group assignments, peer review and edit each other's work, collaborate, and manage digital portfolios. Lapadat et al. (2010) observed that students who used blogs as part of their regular classroom activity were "engaged in their class and eager to read and write online and to collaborate with their peers" (p. 63). Additionally, blogs provided a way to assess a variety of competencies including digital literacy (Lapadat et al., 2010).

Putting Connectivism into Practice

Incorporating blogs into the classroom and having students contribute new information through blogs has been one strategy for applying connectivism. Siemens (2005) identified eight guiding principles of connectivism:

> (1) learning is founded on a diversity of opinions, (2) learning is a process of connecting specialized information sources, (3) technology can aid in learning, (4) being able to acquire and process new knowledge is important, (5) maintaining connections is important for continual learning, (6) the ability to see connections between different concepts is an essential skill, (7) up-to-date knowledge is the focus of all connectivism learning activities, and (8) decision-making is a learning process. (p. 4)

Within the connectivism framework, the starting point of learning occurred when the learner connected and fed information into a community (Kop & Hill, 2008). Kop and Hill provided this summary:

> The learning process is cyclical, in that learners will connect to a network to share and find new information, will modify their beliefs on the basis of new learning, and will connect to a network to share these realizations and find information once more. (2008, p. 2)

Connectivism required learners to be active participants in the learning process with one another and with regard to resources available for knowledge generation (Kop & Fournier, 2010). And as Bell (2010) noted, learning would occur both in and out of the classroom. Integrating collaborative communication technology, such as blogs, in the classroom would allow students to actively share their collective experiences to enrich the learning process.

In this chapter, the authors discussed how the use of a class blog was integrated, as a pedagogical method, into a graduate-level consumer insights class. Investigation focused on whether, as Kop (2011) suggested, access to online networks of information might serve in this scenario as a valuable tool in the connectivism pedagogy.

METHOD

The use of a class blog to create a connected learning environment was included as part of a master's-level consumer insights class available to students in advertising, public relations, and marketing at a large university in the southeastern United States. The course allowed students to be involved in the development of strategic insights as well as strategic decision-making based on insights. The syllabus informed students that upon completing the course, they would be able to uncover important consumer insights that were based in both data and discussion, and that they would learn how to use insights to create successful advertising and corporate marketing communication. The course covered how consumer insights are discovered through examining secondary data and value systems, as well as the development and execution of brand health tracking systems, segmentation studies, and various forms of ethnographic and discussion-based qualitative work. All materials covered and projects assigned in the class emphasized the role of consumer insights in branding, campaign testing, broader innovation, product development work, and more.

For this chapter, the authors used a class blog as a mechanism for creating a networked learning environment where students could share new information they acquired as well as learn from the experiences of others. Through the use of the class blog as a tool for building a learning community, the eight guiding principles by Siemens (2005) were put into practice. By contributing to the class blog (principle 3), students were able to offer their opinions and experiences with the assignment (principle 1), as well as to make connections between their own experience and the experiences of others (principles 2 and 4). The act of updating the class blog throughout the duration of the assignment (principle 7) allowed students to build on existing knowledge (principles 5 and 6), which ultimately aided them in the completion of the final assignment (principle 8).

If, as suggested above, the use of blogs in the classroom not only could enhance critical and analytical thinking, but also could promote creative thinking and collaboration, the potential for blogs to be useful in the course described here was clear. Thus, to better understand the usefulness of connectivism as a concept and blogging as a tool, Regina Lewis (one of the authors) created a two-part assignment in which student blogs were created between completion of a first-part assignment and a second-part assignment to assess whether the presence of a community of learning resulted in improved student performance. A helpful way to think about the method employed in this study might be to consider a pretest (pre-blog availability) versus posttest (post-blog availability) context.

Students were asked to complete a first-part assignment in which they were required to discover little-known but strategically useful information about a mix of goods and services companies. Then, students who had worked to obtain such information created blogs about the product and service categories they had investigated. Contributors to blogs not only posted knowledge that they had independently collected but also provided advice as to where additional information about the various product and service categories might be found. Each category blog was created using WordPress and was posted on the Internet. These blogs served as communities of learning. The overarching URL for all blog content was made identifiable by inclusion of the instructor's name as well as the phrase "class insights" preceding the suffix wordpress.com.

As a second-part assignment, all students were asked to complete parallel assignments in which they obtained strategically useful information about a product or service category that was entirely new to them. In this case, however, students had available to them all of the category blogs that had been created by other students after the first-part assignment; they could glean information from the relevant class blog relating to their new category. When students' performance on the second-part assignment was assessed, rubric-based grades indicated an increase in the value of the information they had gathered. While other factors likely contributed to the higher grades observed, as well, the student-created blogs appeared to have produced positive results.

APPLICATIONS

A Connectivism-Based Course Experience

The two parts of the assignment used in the classroom to examine the blogs' impact were parallel; the only difference between the first-part assignment and the second-part assignment was the client each student was directed to investigate and the presence of a blog to assist in the students' investigation. The specific process was as follows: The class consisted of 32 students, half of whom were communication students and half of whom were marketing students. This balance was maintained for this particular course due to the value of the differing communication and marketing points of view and approaches to problem solving. Thirty-two different client companies were individually matched with students during February 2014. The client companies were carefully chosen to ensure that a wide-ranging mix of service, retail, and not-for-profit companies were used in the study; specifically, quick-service restaurant, hospitality and travel, nonprofit, consumer goods, entertainment, education, sports, and retail clients were included (see Table 2.1 for a complete client/company listing).

Table 2.1. Full List of Categories and Clients Assigned

Quick-Service Restaurants	**Hospitality and Travel**
Chick-fil-A	Gulf Shores/Orange Beach Tourism
Krispy Kreme	Delta
Five Guys	Carnival Cruise
Panera	La Quinta
Nonprofit	**Consumer Goods**
AARP	Warby Parker
ASPCA	Clarks
Salvation Army	Burberry
Red Cross	Louis Vuitton
Entertainment	**Education**
Disney	University of Phoenix
Harrah's	Rosetta Stone
Dollywood	History Channel
20th Century Fox	Harvard Business School
Sports	**Retail**
US Open	Belk
Miami Heat	Publix
X Games	Jos A Banks
NASCAR	Best Buy

In the agency world, in particular, account planners may be asked to understand and act as strategic planners for myriad types of consumers and markets, and consumers and markets of the realms listed above share distinct similarities and differences that merit study. Also, each of the types of companies used in the classroom assignments represented solid potential for growth given current trends and consumer value systems. The National Restaurant Association has predicted continued expansion due to job growth and a resulting lack of time for meal preparation (Restaurant, 2014). The Global Business Travel Association predicted 6.5% growth in 2015 due to improved business confidence and expectations for economic growth (Tilstone, 2014). Reuters expected sports revenue to grow by about 3.7% to $145.3 billion by 2015 due in part to an increasing middle class in key parts of the global economy who readily buy tickets and watch television sports (Stutchburg, 2011). Growth in the for-profit education sector is estimated at 4.1% over the next five years, with technology and results-oriented services expected to be growth drivers (Silber, 2014). Rising online distribution of content and video games should stimulate end-user spending, and factors such as consumer income levels, advertiser spending, technological innovations, and government regulations are expected to impact industry dynamics. On the entertainment front, "The growing dominance of Internet-based entertainment is anticipated to lead to double-digit growth in Internet advertising and account for 10% of overall industry growth during the next six years" (Lucintel, 2012, para. 4). Ernst and Young has said that "the entertainment industry is in the midst of a 'frantic shift' to jump onboard with digital media" (ZDNet, 2012, para. 1). Retail growth, especially online retail growth, was anticipated to grow notably by 2015 (The future, 2015). Obviously, all of these industries, plus the other industries chosen for this exercise, could be important to our current students as they move into professional roles in the workplace.

Once the ideal list of 32 potential clients was finalized, each student was asked to put herself or himself in the role of strategic insights director for a major marketing communication firm; in the scenario, their agencies had just won the accounts with which students were matched. Students were given five days to immerse themselves in their assigned clients' worlds and then deliver "Discovery" briefs for agency senior management that contained "original and interesting" insights as well as "provocative" points about the companies and their industries (see Assignment 2.1 for full assignment document). Students delivered Discovery briefs for the first client assigned to them on February 11, 2014.

Discovery assignments were graded based on a straightforward rubric deemed appropriate for application to the strategic brief assignment. Student work was judged based on the following six components, as reflected in the assignment with which they had been tasked:

1. depth of category understanding;
2. depth of consumer understanding;
3. assessment of the communication baseline (communication to which consumers had been exposed by former agencies up to the current point);
4. customer values and consumer and industry trends coverage;
5. originality of the insights generated; and
6. potential inspiration and excitement generated by the brief.

Three levels of performance for each component were included in the scoring tool, as follows:

1. You needed a bit of coaching on the role of a planner.
2. You're viewed as competent, if not particularly progressive.
3. You've made a great impression . . . This person gets it!

These components and evaluation levels represented typical evaluation tiers in the "real world" of agency account planning document review. Agencies needed (expect strategic planners) to understand challenges that brands have been facing in their categories; to understand often-complex relationships between brands and their consumers; to understand what prior communication has set up current consumer knowledge, expectations, and attitudes (and how); to understand societal trends and values that might be driving these perceptions; and, most importantly, to understand what communication could inspire consumers to desire a relationship with a brand. Lewis recorded all scores in an Excel worksheet. Each student received an individual, overall grade for his or her work, but also received individual rubric scores.

During March 2014, after the first-part assignment had been evaluated, all classroom students performed an in-class, blog creation exercise using WordPress. Because the list of 32 clients used in the first assignment was created using eight different product and service categories (quick-service restaurant, hospitality and travel, nonprofit, consumer goods, entertainment, education, sports, and retail), it was possible to group together students who had completed Discovery briefs for quick-service restaurants (Chick-fil-A, Krispy Kreme, Five Guys, and Panera), for hospitality and travel companies (Gulf Shores/Orange Beach Tourism, Delta, Carnival Cruise, and LaQuinta), for education (University of Phoenix, Rosetta Stone, History Channel, and Harvard Business School), and so on as shown in Table 2.1. This yielded eight groups composed of four students each. In their small groups, students were asked to create category-specific blogs using resources they had used during their work on their first independent Discovery assignments and also resources they might suggest as they brainstormed all of the potential sources of insights they could have investigated and utilized yet had overlooked or not used due to assignment time constraints.

Content for each of the eight category-specific blogs was uploaded to wordpress.com at the end of the in-class exercise, and students were encouraged to continue to add to these blogs as additional ideas came to their minds as time and the class progressed. They were told at the time of their blogging activity and when their blogs were initially uploaded that a second Discovery brief client would be assigned to each class member within a couple of weeks, and that the pool of possible clients would contain the same 32 companies as before. They were also informed that no student would receive a client within the same service or goods category that he or she had previously researched; they were told that each student would effectively be starting from scratch with regard to understanding of his or her client's (1) category, (2) consumer, (3) communication baseline (communication to which consumers had been exposed by former agencies up to the current point), and (4) customer values and consumer and industry trends coverage. They were told that again, students would be evaluated on the originality of the insights they generated and the potential inspiration and excitement generated by their briefs. It was emphasized to the class that given the assignment structure, each student's ongoing blogging efforts would serve to assist other students in the class by providing valuable information sources once the student/client match-ups for the second assignment were announced.

In April 2014, the second Discovery brief was assigned. This assignment was due and submitted April 7, 2014. With the exception of the availability of the class-created industry blogs, the process was identical to the process followed for the first-part assignment in February; each student had five days to prepare his or her Discovery brief. The second-part assignment was graded using the same rubric employed for the first assignment. Again, Lewis recorded all assignment grades using an Excel spreadsheet, but in random order and without reviewing rubric scores that students had received for their initial Discovery briefing work. Again, each student received an individual overall grade for his or her work, but individual rubric scores also were shared.

Observed Performance Outcomes

For all grading purposes, standard rubric technology (www.roobrix.com) was used to convert rubric scores into reportable grades. While this technology is only one of several ways in which rubric conversion can be accomplished (Quinlan, 2012), it was the approach used consistently across the assignments reported in this project. After all rubric scores were converted, student grades for the first assignment ranged from a low of 60 to a high of 90, and grades for the second assignment ranged from a low of 70 to a high of 100.

Of the 32 students who participated in the two assignments, 66% received higher overall grades on the second assignments than on the first; they received higher scores for Discovery work when they had access to online shared social media–based learning about their product or service's broader category. Twenty-two percent of the students received lower overall scores on the second assignment than on the first, and 12% received the exact same grade on the first and second assignments.

Among the two-thirds of the class for whom overall grade performance was higher on the second assignment than on the first, the average positive percentage change was 12%. Positive percentage changes ranged from 30% (the highest) to 4% (the lowest).

Among the approximately one-fifth of the class for whom overall grade performance was lower for the second assignment than for the first, the average negative percentage change was 10%. Negative percent changes ranged from 16% to 4%.

A deeper look at the detailed rubric scores indicated that among those 21 students whose grades were higher on the second assignment than on the first, the most typical area of improvement within the rubric was consumer understanding; three-quarter of those students who showed improvement across assignments also demonstrated improvement on this particular dimension.

A critique of student blog content parallels the finding above, as the blogs that students created after completing their first assignments tended to contain tips, lists of platforms, and suggestions for understanding various category enthusiasts and broader target consumer markets. An illustration of such content was the following, which was pulled directly from the student-created *sports industry* blog:

> We found social media and fan forum websites to be useful. Also, pages to gather information on how fans perceived the sport they support and the opinions they had on it were of help to us. Instagram was helpful when wanting to discover a player's take [on] and perceptions [of] the sport and fans. We also discovered that consumers could be fans of just specific players as opposed to a brand, and this could have a huge impact on whether a fan continued to support the team or organization their favorite player was a part of. It was important to look at the fans who followed players, and why they felt connected to that certain player and admired them.

This same four-person group suggested a list of platforms for insights into the sports industry. The list was quite consumer-heavy in nature. Group members suggested that anyone investigating this industry explore Facebook; Twitter; Instagram (players' profiles); official websites; articles by fans; YouTube; fan forum (to discuss their common interests); Hoovers; Foursquare; and blogs.

The four-person class group that created the *education industry* blog suggested several broad industry links but also focused the bulk of their blog content on consumers, in particular. Their advice to students who would investigate this category during the second round of assignments included the following bulleted points:

- Look on LinkedIn to see if the university students have been speaking the language of their industry or seem to be prepared to work in their field.
- Follow specific students or alumni, heavy users or enthusiastic fans on social media to understand more of the consumer's view on the educational resource (universities, History Channel, and Rosetta Stone).
- Find heavy users of Rosetta Stone or other educational resource and interview them on their experience and how they found out about the product.

- Try to find more creative research [methods] such as . . . asking people on Twitter or Facebook what their perceptions were of the brand.
- If possible, find graduates and non-graduates, users and non-users, and try to interview or survey them about the education tool or university.
- Check Pinterest for more followers or fans.

Finally, blog content created by the four students who initially wrote their Discovery brief for a *mock hospitality* or *travel client* included the advice and observations shown in bullets below:

- If possible, try to go outside of the company and speak to consumers or friends who have taken recent trips on what they liked and didn't like.
- Follow social media hashtags and pages for a few days to get a sense of feedback from consumers.
- Check feedback/comments/comparisons of companies on websites like kayak.com.
- Talk with people traveling for Spring Break and other trips and [see] how they decided where to stay, who to fly with, etc.

A deeper look into scores for each of the rubric components also indicated that the majority of students whose grades were higher on the second assignment than on the first showed improvement on the values and trends dimension. Thirteen students showed improvement on this dimension, specifically.

As was the case with the consumer understanding dimension, an examination of student-created blog content revealed information definitely related to the trends and values dimension, specifically. Several of the student groups blogged about industry trend sources, in particular, and provided relevant links to Hoovers' trend site and others. The entertainment industry group included a section on their blog titled "Entertainment Trends We'll Applaud in 2014" and provided multiple links to entertainment trend information for those students who would be assigned entertainment clients later in the term.

DISCUSSION

This exercise applied the theory of connectivism within the context of a master's class composed of advertising, public relations, and marketing students. The overarching goal of the authors was to better understand ways in which information shared in the classroom in the form of social media might enhance performance and learning. More specifically and immediately, the goal of the authors was to improve student performance on an open-ended assignment in which student-contributed information was made available via blogs to provide a starting point and spark new ideas among aspiring strategic communications students. Because the communication and marketing fields require innovative thinking yet demand that high-pressure deadlines be met, the consumer insights and understanding class examined in this work was deemed conducive to implementing innovative teaching methods using social media. The consumer landscape has ever been changing, meeting the criterion of a rapidly changing knowledge foundation on which the theory of connectivism was partially based. The authors embarked on this project with the belief that if it appeared true in this course that students could benefit from being active participants in learning, then that knowledge could be extended moving forward and likely be relevant to educators offering broader communication campaigns and marketing planning courses. The core use of social media (blogs) in this assignment could spur similar uses of social media across many courses involving campaign development.

The importance of this blogging exercise to students could be justified by noting that the role of social media in future agency campaign success cannot be minimized. Incorporating social media into the classroom and in the particular way that was chosen for this exercise seems especially critical. Blogs do, indeed, promote creative and associational thinking, promote analogical thinking, lead planners toward quality information, and encourage social interaction. And, in looking back at Siemen's eight guiding principles of connectivism, associations between what would be required of agency planners and how connectivism could provide value to them became clear. If connectivism indeed is founded on (and, therefore, fosters) a diversity of opinions, could aid in learning, could help people process new forms of learning, and could illuminate connections between concepts while keeping researchers up-to-date, the theory has been essential to the communication and marketing fields.

The ramifications of positive experiences with connectivism in the classroom certainly could extend beyond communication and business programs. Because connectivism has been a theory of knowledge management (Siemens, 2005), and knowledge sources continue to widen, it would be increasingly important to understand ways of collecting, maintaining, and building on information for students. Particularly in a master's-level environment, it seemed both efficient and effective to share some of the burden of keeping knowledge current with the students. While the exercise covered in this chapter focused on ways in which the use of student blogs could improve student performance on an assignment, it also could be assumed that the practice could improve student preparedness for the corporate world. In that world, they would be asked to quickly discover and distill large amounts of information and tailor presentation of that information to their clients' multiple needs.

The authors certainly could identify some limitations of the exercise conducted in the present work. First, the authors acknowledge that other factors likely influenced students' propensity to earn higher grades on the second assignment than they earned on the first. The students were more likely comfortable with the style of the assignment and the content of the rubric in April than when first exposed to these documents in February, and they had likely discussed the first assignment among themselves after receiving their initial grades.

Also, it was interesting to consider the areas of rubric improvement observed in April. In any ongoing project, the consumer insights dimension has been always challenging; the same would be true of the ever-changing values and trends dimension. Discovery within both of these areas required the use of unique and creative information sources. Perhaps, sources shared in the class blogs helped students overcome hurdles in obtaining knowledge. Perhaps, the obtaining of knowledge became less intimidating to students. Certainly, information that could have been helpful to students existed within the body of the blogs that class members shared.

An interesting next step in understanding the effectiveness of group blogging in the classroom context would be the extension of the blogging timeframe. In the exercise conducted and reported upon here, the "lifetime" of the blogs was only one month; in light of constantly changing knowledge foundations and sources, as well as the connectivism principles that related to (1) acquiring and processing new information and (2) making connections between new knowledge and already existing knowledge, longitudinal research would be appropriate.

Regardless of an instructor's approach to enhancing learning through new media, several aspects of classrooms and their possible outcomes today appeared indisputable. First, digital technology appeared to have claimed the heart of many of the global industries in which our students intend to succeed. Whether students' aspirations center around sports, entertainment, not-for-profit, quick-service restaurants, travel, or consumer retail, they would have little hope of success unless equipped not only with a clear understanding of the business implications and uses of digital tools, but also with an arsenal of digital knowledge with which to communicate and market to consumers and other shareholders alike.

And, once communication and marketing enter the business picture, social media play an irreplaceable role. It is through social media that the many players within and outside of organizations discuss, debate, and further their industry growth; social media is the current form of information dissemination that is most preferred by many in the world in which our students hope to successfully enter. While blogs, the form of social media that received focus here, represent only one type of social media approach, it would appear that they remain important for educators to keep in mind moving forward. If blogs can not only help students build credible corporate identities but also allow them to amass, incorporate, and make sense of ongoing, massive amounts of information and data that are likely to come their way, it is more important than ever before that students be adroit in creating, managing, and maximizing their potential.

Connectivism is a knowledge management theory that organizations are likely to appreciate as they move into more complex global and technological spaces; it is interesting that this theory can be applied in the classroom at a basic level as well as demonstrating almost unlimited possibilities for enhanced communication in varying spheres. Perhaps because of its flexibility, in addition to its demonstrated value in the exercise described in this chapter, it should become a classroom staple as forms of digital learning methods continue to be explored.

It is the goal of instructors, with few exceptions, to prepare students for success in an increasingly complex business world. Given this, it is hard to imagine a classroom that fails to experiment with social media in many forms in a quest for excellence.

REFERENCES

Bell, F. (2010). Network theories for technology-enabled learning and social change: Connectivism and actor network theory. In *Proceedings of the 7th International Conference on Networked Learning*, pp. 526–533.

Churchill, D. (2009). Educational applications of Web 2.0: Using blogs to support teaching and learning. *British Journal of Educational Technology*, *40*(1), 179–183.

Duffy, P. D., & Bruns, A. (2006). The use of blogs, wikis and RSS in education: A conversation of possibilities. In *Proceedings Online Learning and Teaching Conference 2006*, pp. 31–38. Brisbane, Australia.

Higdon, J., & Topaz, C. (2009). Blogs and wikis as instructional tools: A social software adaptation of just-in-time teaching. *College Teaching*, *57*(2), 105–110.

Kop, R. (2011). The challenges to connectivism-based learning on open online networks: Learning experiences during a massive open online course. *The International Review of Research in Open and Distance Learning, Special Issue-Connectivism: Design and Delivery of Social Networked Learning*, *12*(3).

Kop, R., & Fournier, H. (2010). New dimensions to self-directed learning in an open networked learning environment. *International Journal of Self-Directed Learning*, *7*(2), 2–20.

Kop, R., & Hill, A. (2008). Connectivism: Learning theory of the future or vestige of the past? *International Review of Research in Open & Distance Learning*, *9*(3).

Lapadat, J. C., Brown, W. I., Thielmann, G., & McGregor, C. E. (2010). Teaching with blogs: A case study of technologically mediated literacy. *International Journal of Learning and Media*, *2*(3), 63–79.

Lucintel. (2012). Lucintel's research indicates global media and entertainment industry to reach US $1,289 Billion in 2017. Retrieved from http://www.lucintel.com/press/global_media_entertainment_industry_2012_2017.aspx

Mak, S., Williams, R., & Mackness, J. (2010). Blogs and forums as communication and learning tools in a MOOC. In *Networked Learning Conference* (pp. 275–285). Lancaster, UK: University of Lancaster.

Martindale, T., & Wiley, D. A. (2004). Using weblogs in scholarship and teaching. *TechTrends*, *49*(2), 55–61.

Quinlan, A. M. (2012). *A complete guide to rubrics: Assessment made easy for teachers of K–college* (2nd ed.). Plymouth, UK: Rowman & Littlefield Education.

Restaurant industry forecast 2014. Retrieved from https://search.yahoo.com/yhs/search?hspart=GenieoYaho&hsimp=yhs-fh_ds&type=a19400001&p=nra+forecasts

Siemens, G. (2005). Connectivism: A learning theory for the digital age. *International Journal of Instructional Technology and Distance Learning, 2*(1), 3–10.

Siemens, G. (2006). *Connectivism: Learning theory or pastime of the self-amused.* Manitoba, Canada: Learning Technologies Centre.

Silber, J. (2014, September 1). Forecasts for growth in the education industry. Retrieved from http://www.educationindustry.org/industry-trendsdata

Stutchburg, G. (2011, December 9). Sports industry expected to continue growth despite economic woes: Report. Retrieved from http://www.huffingtonpost.com/2011/12/09/sports-industry-expected-grow-2015-report_n1138706.html

The future of retailing in the United States to 2015. Retrieved from http://www.marketresearch.com/iCD-Research-v3849/Future-Retailing-United-States-6424699/

Tilstone, P. (2014, March 13). GBTA forecasts year-over-year business travel growth in Europe 2014, 2015. Retrieved from http://www.gbta.org/foundation/pressreleases/Pages/rls_031314.aspx

ZDNet. (2012). *Ernst & Young: Digital entertainment revenue to surpass "traditional" media by 2015.* Retrieved from http://www.zdnet.com/article/ernst-young-digital-entertainment-revenue-to-surpass-traditional-media-by-2015/

ASSIGNMENT 2.1. PROJECT DISCOVERY: FIVE DAYS IN THE LIFE OF AN ACCOUNT PLANNER

You are an Account Planner/Strategic Insights Director for a major marketing communications firm. Your team learned today that you have been awarded the Krispy Kreme account! This carries a significant communication budget and will result in millions of dollars of revenue for your agency. This is your team's first foray into the world of quick-service restaurants.

Your first step in overseeing Insights for this account is DISCOVERY. In this initial phase, you are trying to gather insight and inspiration. You should aim to explore as broadly as you can, to make the most original and interesting discoveries.

What are some discovery places? Of course, an understanding of the "fanbase"! This likely involves following social media, along with any other ways in which you can understand—in the timeframe you have been given—what your loyal consumer may think and feel. What does your brand help people do? You also want to understand recent brand communication (you have "stolen" the account from what has been the current agency, and want to understand the baseline communication to which you must be superior).

You will need to discover information about the category. What do other brands do and what can we learn from that? Also, how might they deal with the same problems that Krispy Kreme faces? Finally, who are the category innovators?

As we have discussed at length in this course, relevant larger value systems and social trends will play into your discovery phase. What values in life are your brand associated with, and what trends could play into its success?

The main purpose of the Discovery Report is to summarize the most salient, most provocative points that have arisen from your first five days of immersing yourself in the Krispy Kreme brand. Your task for Tuesday (February 11) is to share your Discovery Report to date with your senior leadership, who are very hungry to understand the "situation analysis" of the new client for whom they intend to knock communication efforts out of the ballpark.

In the end (after your report to your senior management), your Discovery Report will be no more than two or three pages. Initially, however, it may be up to five pages long. This is the document you will write for February 11.

We will discuss your Discovery Report in our next classroom session. Until then, I encourage you to "live and breathe" your new client brand! I think you could be surprised at the revelations that could come to you.

We won't have formal presentation on February 11, just solid and interesting discussion. Each of you has a different new "client," and so you will be working individually.

I look forward to seeing what you DISCOVER!

GOOD LUCK!

DR. LEWIS

CHAPTER THREE

Being Your Own Chief Marketing Officer

Student Perceptions of Personal Branding

HOLLY KATHLEEN HALL AND MARY JACKSON PITTS

Brand positioning is defined as "the active communication of one's brand identity to a specific target market" (Labrecque, Markos, & Milne, 2011, p. 44). Business organizations need a well-managed, readily identifiable brand to sell products and services. Branding is no longer just for products. Personal branding, emerging in the 1990s, functions much like company brand positioning, communicating a certain image and building relationships (Lair, Sullivan, & Cheney, 2005). According to Peters (1997), personal branding urges everyone to use the same concepts taken from product branding and apply them to ourselves:

> That cross-trainer you're wearing—one look at the distinctive swoosh on the side tells everyone who's got you branded. That coffee travel mug you're carrying—ah, you're a Starbucks woman! Your T-shirt with the distinctive Champion "C" on the sleeve, the blue jeans with the prominent Levi's rivets, the watch with the hey-this-certifies-I-made-it icon on the face, your fountain pen with the maker's symbol crafted into the end . . . . You're branded, branded, branded, branded. It's time for me—and you—to take a lesson from the big brands, a lesson that's true for anyone who's interested in what it takes to stand out and prosper in the new world of work. (para. 1–3)

A brand is a perception. Building a personal brand involves developing the image that you want to portray to influence how people perceive you. Personal branding has also been described as "engaging in self-commodification" or "turning oneself into a product" (Lair et al., 2005, p. 319). According to Goldsmith (2009), crafting a personal brand is

> an evolutionary and organic process that should emerge from your search for your identity and meaning in life . . . . It's the key to personal success . . . as it will help you distinguish yourself as an exceptional professional. (para. 4–5)

A personal brand can influence an employer when they make a hiring decision. It is, therefore, crucial that today's graduates know the importance of building a reputable personal brand for their professional careers. While today's students may be perceived as digital natives or "digital sophisticates," students have social media knowledge in that they know how to use social media tools, but they lack the understanding of using it strategically, nor do they appreciate the consequences of the digital footprint they may be constructing (Wetsch, 2012, p. 30). Increasingly, employers have been using the Internet and social media as a method for investigating potential employees in deciding whether to hire or even interview a candidate. Seventy-five percent of hiring managers have investigated possible hires through such methods (Dutta, 2010). Social media sites such as LinkedIn and Facebook are specifically being used by employers to get a sense of a job candidate's character and personality, along with personal websites, as it is "easier and safer for them to gain insight to who you are before offering in-person meetings to explore what you are" (Morgan, 2011). Websites are hailed as a particularly effective method for communicating a personal brand as 56% of hiring managers are more fascinated by a candidate's website than any other social media tool, yet only 7% of potential employees reported having a personal website as of 2013 (Smith, 2013).

Employment trends in terms of how long a person stays with one organization have changed as well. The era of staying with one company an entire career is over. To survive a competitive, constantly changing professional world, it behooves the employee to be unique, noticeable, and entrepreneurial, demonstrating the desire to add value, "taking control of your own success" (Lair et al., 2005, p. 321).

Yet another reason to focus on a personal brand is that those brands can be leveraged to bolster an organization's already existing brand. As Arruda (2013) stated, "many companies focus all their branding efforts on marketing activities such as advertising campaigns and packaging, yet one of the most powerful brand assets your company has is your people" (para. 2). Every interaction a customer has with an employee is an opportunity to positively reinforce the company brand.

The purpose of this chapter is to provide tools instructors may use to help their students in developing their personal brands and to provide information regarding student perceptions of personal branding so that curriculum can be carefully tailored to match the beliefs and expectations students have regarding personal branding. This study used an online combined senior undergraduate and graduate social media class and a senior undergraduate brick-and-mortar class at a southern regional university to examine perceptions of personal branding among the students and included assignments to assist students in discovering, developing, and refining their personal brand.

LITERATURE REVIEW

Having a positively perceived personal brand depends on several factors such as the level of authenticity and knowing what kind of content would potentially be brand-damaging. For example, typos, outdated information, and the inclusion of controversial topics such as certain political or religious posts can be detrimental (Smith, 2013). As Burnside (2014) stated, "know what you are aiming to do and who you're keen to impress; self-promote like a boss; try not to leave a trail of devastation in your wake that could backfire on you" (para. 5). Marwick and Boyd (2010) found revealing carefully selected, non-controversial personal information is "a marker of authenticity" (p. 127). However, users are also cautioned to post true, honest personal information and not present an inauthentic, false, or overly polished self (such as using too many noticeably staged/posed photographs) (Labrecque et al., 2011). Some take the opportunity to inappropriately embellish their résumés, yet personal branding "relies on the language of authenticity, arguing the responsible self-brander is a person who is honest with herself and others" (Gehl, 2011, para. 79).

While there have been benefits to a strong personal brand, there have also been risks. Company leaders and administrators would fear employees who might violate organizational, governmental, or industry rules while using their personal brands to interact with customers, as well as infringing on someone else's intellectual property (Dutta, 2010). Not only did company leaders feel a loss of control, but the user himself or herself could not completely control the content, and information a user posts can be shared by friends and third-party applications, exposing users to the risk of their content being unknowingly shared or changed (Labrecque et al., 2011).

The audience on social media platforms must also be considered. A person's Facebook, for example, could be an amalgamation of friends, family, co-workers, and casual acquaintances. Even with privacy settings enabled, content is being created and posted to a diverse audience. Users might choose to only "play" to one specific audience among their friends or followers or to appeal to diverse audiences at the same time. Marwick and Boyd (2010) refered to the latter as "polysemy" or using "coded communication to simultaneously appeal to different, even oppositional audiences," mentioning the example of Madonna in her early career as she concurrently appealed to young women who identified with her feminist image and young men who recognized her more for her sex appeal (p. 123). Therefore, communicating the personal brand online has often been a balancing act, as users had to determine what information would be relevant, helpful, and resonant.

The actual process of developing and managing a personal brand would go deeper than the information normally contained in a traditional résumé. It involves a progression of self-discovery to determine values, interests, personality, life goals, and character. Goldsmith (2009) outlines a four-step model for building a personal brand:

> 1. Define and formulate your personal ambition.This means [accessing] your personal vision, mission, and key roles—and make them visible.
> 2. Define and formulate an authentic personal brand promise that you can use as the focal point of your behavior and actions. Your personal brand statement entails the total of your ambition, brand objectives, specialty, service-dominant attribute, and domain.
> 3. Formulate your personal balanced scorecard. . . .
> 4. Implement and cultivate your brand. Personal branding has no value unless you make it a reality. So create and maintain your brand effectively. (para. 8)

Umar (2014) had a five-step plan that he recommended:

> first performing a self-assessment to determine your values, personality, and interests; second, getting feedback from others to see if your description is accurate based on the perceptions other people have of you; third, look for common phrases, themes and descriptors based on the information gleaned from the first two steps; fourth, rank your top five "brand elements"; and finally use that information to craft a personal brand statement, using emotive and unique words such as "Bobby is the Beacon of Light who Helps 'Lost' Leader-Ships ashore to start an Authentic Path of Passion and Purpose." (para. 17)

Once a user decides on the brand image, the next decision is which social media platforms to use. Most human resource professionals cited LinkedIn as a necessary site on which users should have a presence (Langston, 2014). The profile on LinkedIn at a minimum should contain a recent photo, work and educational information, and a clear look at major skills and successes (Morgan, 2011). Depending on what a user wanted a personal brand to do for them, other appropriate tools might be blogs, podcasts, and social networking sites such as Facebook. Some branding experts cited websites as the key tool to use in conveying a personal brand, as it provided "creative freedom to express your personality in ways that are not possible through your resume," offers visibility, and demonstrates the user's determination and care in finding a career (Smith, 2013, para. 4).

When profiles on appropriate sites are established, the task then becomes one of maintaining and monitoring. Dutta (2010) recommended Googling yourself habitually to see what arised and tracking connections each month to see what benefits might be emerging due to social media interactions. Labrecque et al. (2011) suggested getting regular feedback on social media posts and conversations. In Marwick and Boyd's study (2010), users got feedback from close friends asking questions such as "Do I look too much like a marketer?" and "Am I tweeting too personally?" (p. 127). Using this kind of feedback allowed the users to effectively manage their personal branding strategies. Those strategies would and should evolve as social networks evolve and potentially would become unhelpful to a personal brand or as people's careers and lives progress (Labrecque et al., 2011). As Wetsch (2012) noted,

> if social media knowledge and intelligence were taught in an engaged academic setting, before individuals enter the work force, it would be possible to bridge the gap between schooling and social media that many of today's graduates encounter in the "real world." (p. 32)

Branding would bring to mind products and services. And the phrase "set yourselves apart from others" has been used to help job hunters. Branding of oneself has commonly been applied to thoughts of making sure one's hard-copy résumé looked good and one's portfolio showed work in a formidable way. However, in our electronic social world, students have been creating a social media brand each time they post; they unsuspectingly have been creating their "personal brand." With this in mind, the researchers sought to gain insight into the following questions:

RQ1: How did students define personal social media branding?

RQ2: Why did students believe personal social media branding was important?

RQ3: What were students' behaviors toward personal social media branding?

RQ4: Were students' professors addressing personal social media branding?

METHOD

The study examined university students' perceptions of personal branding in relation to the research questions above. Upon analysis of data, the researchers proposed several class assignments, which were used (1) to create awareness of one's personal brand and (2) to provide avenues for developing their personal brand based upon a SWOT analysis, market analysis, and business plan while focusing heavily on the public relations, marketing, and advertising of their core personal brand.

A sample of students at a southern, regional university were asked in an email to complete a 71-item survey in the summer of 2014. Students were provided with an incentive for answering the survey: the potential to win one of four $25 gift cards to a local store. The original email link remained active for five days after it was sent on June 2, 2014, and the data collection ended on June 6, 2014. Data collected from the survey were analyzed using Statistical Package for the Social Sciences (SPSS). Of those responding, 45 students were communication majors. It is the group of 45 who became the focus of this study on personal branding in social media.

Respondents' demographics data and data regarding respondents' top three social media sites, number of social media sites maintained, and the audience for respondents' social media content were collected. Additional variables addressed how students defined personal branding and reasons personal branding has been important. In addition to several closed-ended questions, students were asked several open-ended questions that allowed them to write about their own personal definition of social media branding and why social media branding would be important.

Several Likert statements were used to measure students' behaviors toward their own personal social media brand and the tools that were useful in managing their social media reach. These included items related to presence in social media, whether blogs were used, if they had respondents to request to remove social media content, and other items related to level of professionalism with their social media presence. Respondents were also asked using an open-ended request to describe their personal brand and to rate that brand. An additional focus was placed on what information students receive from professors about personal social media branding.

APPLICATIONS

Forty-five communication students who were focusing on advertising, public relations, creative media, multimedia journalism, and communications studies agreed to participate in the study. Respondents were closely aligned with the universities' demographic make-up. The mean age was 25. Sixty-four percent (n = 29) were female and 35% (n = 16) were male. The mean number of hours completed was 68. The majority of the respondents (66.7%) were Caucasian; 13.3% were black or African American, 4.4% were Hispanic or Latin, 4.4% were Asian or Pacific Islander, and 4.4% were American or Alaskan Native. The remainder, 6.6%, preferred not to answer the question.

When asked to list the top social media sites, Facebook was the top site identified (51.1%) followed by Instagram (17.8%), Twitter (17.4%), and LinkedIn (2.2%). When asked to identify the second most important social media site, Twitter was the top second choice (22%), followed by Instagram (20%), LinkedIn (13.3%), Facebook (11.1%), Pinterest (11.1%), and YouTube (11.1%). Twitter took third place as well as the third most go-to social media site with 24.4%, followed by Facebook (17.8%), Instagram (15.5%), LinkedIn (11.1%), and Snapchat (6.7%).

Communication students maintained 4.13 social media sites and indicated that friends were their number-one audience followed by school peers, work peers, family, potential employers, entertainment peers, church, and the worldwide audience.

RQ1: How Did Students Define Personal Social Media Branding? When presented with a definition of personal branding, most respondents (95%) indicated agreement with the definition that personal branding is an individual selling themselves rather than a company or brand ($x2=36.36$, df=1, p <.001). Ninety-three percent of the respondents agreed with the statement that personal branding will help one get a job (x2=33.8, df=1, p <.001).

When students were asked to define personal branding, a multitude of definitions were provided with each giving a glimpse of a small picture of each respondent's personal branding definition.

RQ2: Why Students Believed Personal Social Media Branding Was Important? Educators can be guided by their answers as they develop curriculum to prepare students for the workforce. Examples of students replies were "it is important because a solid personal brand can give you that slight edge over another person who you are competing with for a position"; "it is a way to get yourself looked at and to set yourself apart from everyone else"; "in today's competitive job market it is a tool that is necessary to be noticed"; and so forth.

RQ3: What Were Students' Behaviors Toward Personal Social Media Branding? Sixty-four percent of respondents thought they had a brand and agreed that they had a solid personal brand among the social media platforms. Forty percent believed their brand was professional. Eighty-four percent indicated they try to make sure their personal profile description was consistent across social media platforms (x2=21.35, df=1, p<.05). The majority of students (75%) did not brand themselves using a personal blog (x2=5, df=1, p<.05). Eighty-nine percent used their given name as their profile name (x2=27.22, df=1, p<.001) and 82% thought their profile name was professional (x2=18.68, df=1, p<.001). The majority (73%) believed their brand could easily be found by potential employers (x2=9.80, df=1, p<.002); 89% indicated they keep their social media platforms current (x2=27.22, df=1, p<.001); and most (71.7%) post similar content across all platforms (x2=3.75, df=1, p<.05).

Sixty-seven percent of the students reported they did *not* use their personal blog to focus on their strongest skill set (x2=5.00, df=1, p<.01); advance their careers (67%) (x2=5.00, df=1, p<.01); nor did they have their own (73%) domain name (x2=9.8, df=1, p<.002). Students (69%) had removed photographs from social platforms that might appear unprofessional (x2=6.4, df=1, p<.01); and most (91%) did have a résumé on file that they would update regularly (x2=30.42, df=1, p<.001). Only 53% used Facebook to promote themselves professionally and a slightly larger number (58%) used LinkedIn. Fewer (18%) used Twitter to promote personal blogs (x2=18.68, df=1, p<.001), and LinkedIn was not being used to promote a personal brand.

Fourteen percent had items in their profile that they would not want a potential employer to see (x2=40.20, df=1, p<.05). Communication students rated their personal brand scale at 3.59 with 5 being the best. And, there was little concern about potential employers seeing their social media content. The mean was 2.00 with 5 being the most concerned.

Knowing what personal branding was and understanding why it was important were variables in establishing one's own personal brand. Asking students to define their own personal brand yielded a wide variety of answers, which showed a need for a more developed curriculum that would allow the students to better explore how they would wish to be seen in the business world (see Assignment 3.1).

RQ4: Were Students' Professors Addressing Personal Social Media Branding? That is to say, how were college professors addressing personal social media branding? Students (77%) reported their teachers were likely to talk about personal branding in their classes (x2=13.88, df=1, p<.001), and students were more likely (84%) to hear about branding in their major than outside of their major (x2=21.35, df=1, p<.001). Few students (24%) were required to have a blog (x2=11.75, df=1, p<.001). Only five students had ever had a professor suggest to them to remove questionable material from their social media presence.

Based on the survey, many students recognize what personal branding was and its importance in finding a job. The actual intentional, purposeful preparation of their personal brands, appeared to be lacking. In particular, blogs and websites as personal branding tools were profoundly underutilized. It appeared professors were addressing personal branding, but few were requiring students to create personal brands by using blogs and personal websites. In addition, more focus on professional social media sites such as LinkedIn would be helpful for these potential graduates. Students reported that friends were their number-one audience and perceived potential employers as their fifth most likely social media audience. This would be a cause for concern.

The takeaway for instructors would be to find or develop assignments that help students hone and further focus on their personal brands. The personal brand descriptors stated by respondents were incomplete and generic. Some responses indicated an insecurity about revealing "who they are." Assignments or projects that foster and encourage deep dives into each student's unique and valuable attributes would assist them in talking confidently about themselves, whether online or in a face-to-face interview. Students needed to be taught that their social media brand would be for the eyes of the potential employer, and creating social media content that reflected a student's work and skill set would be an important start to producing a social media brand that would be interesting and attention grabbing for a potential employer. The following assignments were examples that can be used to help students develop and refine their personal brands.

Assignments

Students needed the awareness, understanding, and skills to develop their own personal brands in an academic setting, which would then position them to assist organizations in developing brands for products/services and with successful customer interactions using online interactions. Students could learn the tools and strategies for personal branding in the classroom that would assist them when they enter the real world.

Assignment: Discovering Your Personal Brand. Through an online class assignment, students in a Social Media in Strategic Communications class were introduced to a method for discovery and development of their personal brand image. The students were asked to spend some time in class reflecting on who they were and what they were about. They were asked a series of questions to get to the core of who they were (see Assignment 3.2).

The students wrote down their responses to those questions and shared their responses with the class. Then they were instructed to distill some of those responses into a two- to three-sentence descriptor such as "I am a mother/teacher/facilitator with a love for learning and my students. I am a writer, creator, researcher, and foodie, and I revert to childlike behavior with my son. I also have a strong desire to instill the importance of ethical behavior in my PR students," using descriptive words that would be unusual, unique, and not excessively used in general, everyday conversation. The students then shared what they had written. To some students, it was an easy exercise. They were confident and sure of themselves and their goals. For others, this exercise was a struggle. Many of them were at that place in their lives where they still were not quite sure "who they are" and what they had to contribute to the world. Sample statements collected from students in a 2014 class include the following:

- I am a teacher/brother/friend. I am a hard worker, ambitious, stylish, adventurous, and love to cook. I am affectionate with children so, I forget about myself when I see my nieces and nephews. I aspire to develop research in different areas of media.
- Through my leadership, intuition, and genuine concern for—and interest—in others, I build long-lasting, fruitful relationships with my team, the citizens I serve, to provide better opportunity, and instill hope in each and every one of their lives.

These students were then encouraged to do additional research such as identifying long-term and short-term goals, performing a SWOT (strengths, weaknesses, opportunities, threats) analysis on themselves, listening actively to words other people use to describe them, to carefully consider who they need their target audience to be, and to contemplate if they had an area of thought-leadership or specific point of view that they are known for or enjoy pondering. Then, they were asked to answer these questions:

- Which tools (social media platforms) will help me the most? Why?
- What is the purpose of the tool, and does it fit with what I need to accomplish?

Once the students felt like they had a solid descriptive statement and they knew the platforms that will assist them the most, they were encouraged to use their descriptive statement on their social media platforms for consistency of their personal brand.

Assignment: Media Management and Entrepreneurship. The media management and entrepreneurship class was designed to help students think beyond the present to what they would become when they graduate from the university. Many aspired to graduate but they feared that day as well. They should find a job working for themselves or for a potential employer. This class was designed to help students create their personal brand by exploiting the knowledge they have gained. Selling their personal brand was the core focus of this activity. This was a senior-level class taken in the last two semesters of a student's college experience. This was a semester-long activity that resulted in the students creating their own personal brand and products. The area of expertise could be within their field of study, their hobby that could become a business, their extracurricular activities, or another product or service that they believe was missing from the commercial market. Students spent the entire semester developing their personal brand based on a SWOT analysis, market analysis, and business plan. Students were required to focus heavily on the public relations, marketing, and advertising of their core personal brand. The following described the stages of the students' semester-long project (see Assignment 3.3).

First Phase. Students were introduced to the changing media climate. Focus was on legacy media and social media outlets. Students look at how audiences have been shifting their behaviors toward media. Data from Edison, Arbitron, and Neilson (2014) were used to show the shift away from traditional media. The data reflected the power and influence of social media for a person and for a business. Discussion focused on how best to reach a potential audience and who the audience was for a different medium. Audience habits and behaviors were addressed as well as buying habits. Students were introduced to why people purchase what they purchase and what needs those purchases met. Both demographic and psychographic data were considered as students began to become more aware of potential audiences and behaviors toward brands and people. Students began to develop an awareness of how a brand would become important to audiences within different media.

Second Phase. The second phase of the semester was a personal search of who the students were and what product or service they believe was missing from the market and how they might be able to address that market deficiency. Students were asked to consider their attributes that can be used to create their personal brand. Students were encouraged to use the expertise that they have developed throughout their college career to find their personal passion and to develop that passion into a money-making personal venture. Small-business leaders, bankers, and professional entrepreneurs were brought in during the second phase of the class to introduce the students to creating a personal focus or brand. Personal branding included a discussion of being a good financial citizen, finding one's personal passion, developing relationships with community leaders and supporters, taking the step forward into being your own personal brand, and developing a strong media presence to include legacy and social media. Creating awareness of the power of social media and the relatively inexpensive nature of social media branding has been critical in a student's development of a social media presence. To that end, a specialist in social media introduced students to analyzing and using social media analytics such as Facebook Insights and Google Analytics to the class. After meeting with these professionals, students were required to create their personal brand and/or product/service that they believe is needed in the market. Many times, students used their home residence as the location of their business/personal brand launch. Students were asked to develop a SWOT analysis of a personal brand or product extensions of that personal brand. There were times when this SWOT analysis resulted in a redirection of the overall personal branding project; students found upon research that they might need to go in a different direction for the branding project to be feasible and lucrative.

Third Phase. For the third phase of the class, the students embarked on a market analysis for their personal brand and any products that extend from their personal brand. Students were encouraged to focus on their personal brand and products of their brand. In this phase, students produced a market analysis whereby they analyzed the cost, potential revenue, brand development, marketing cost, advertising cost, public relations cost, and multimedia and social media costs. At any point in the phases, students were allowed to change their personal branding focus. Once the market analysis was complete, it was likely that what the student thought was the correct branding strategy or product for the market shifts after the analysis revealed flaws in their branding ideas and intentions. In essence, the students were attempting to launch their personal brand. This was a rather uncomfortable phase for students as they wrestled with the reality of who they were and how their brand should appear in social media. For instance, one student progressed through the first three stages only to find that the student was four weeks from graduating, and what had initially been created as the brand was nowhere near what the student's goals and aspirations were for the future. Because of the flexibility of the process, the students were able to redirect and created an excellent personal brand based upon their real personal skill set.

Fourth Phase. The final branding occured through the development of a business plan complete with a social media branding campaign found inside a personal branding website. Students were given the latitude to choose the platform they wished in creating the site. Home pages were used to create a visual representations of each student, complete with information about the individual, the market brand the individual was trying to promote, and the services provided by the individual brand holder.

The following examples were derived from this class project. A female student in the class developed a personal brand that focused on her wide experiences in video work. It took this student two tries to figure out how she wanted to brand herself. In the third phase of the branding assignment, the student, after spending countless hours on a personal brand that did not focus on her strengths, found herself as she began to recognize the overpowering fact that she was graduating with her bachelor's degree within six weeks and needed to set herself apart from others. Thus, she developed a personal website that focused on her greatest strengths in video production. The outcome was a website that focused on her skills to promote businesses in the the marketplace. The website addressed her skill level, her desire to help businesses create stories about themselves, and an example of the type of video work that she was capable of producing. In this exercise, the student used her personal name as part of her web address, thus giving her name and business focus. A second student combined her passion for Christian missions with her degree in radio-television to create a business focusing on capturing the stories of individuals who spend time helping other people. A third student used his graphic design skills as a center point of his website, which provided a showcase of his design work using video, photography, and drawing skills to showcase his work in advertising. A fourth student created a website that showcased his skills in writing, video work, and promotion of the development of an app. In all of the students' work, a résumé, skill set, and contact information were made readily available to the public. Additionally, students provided a mission statement for themselves and the products they promoted.

DISCUSSION

Personal branding should be introduced to students early in their college careers, and the collection of products to showcase student works and experiences should be started during the first semester that a student arrives on campus. Communication programs should encourage faculty to mentor students so that each student would be pleased and proud of their personal and individualistic brand. Data collected in this study indicated that faculty were talking to students about personal branding, yet less than 25% of students indicated that faculty were actually requiring students to create a personal social media brand for themselves. In the past, paper résumés served as the personal branding tool, and the objective of the job search was succinctly stated. Today the employer expectation for communication students is that they have an online presence that showcases the skill set and a portfolio of their work and experiences. Faculty must work to help the students create a personal brand and then actually require their students to showcase their work, motivations, and dreams. The proliferation of online tools, such as Wix, Weebly, and WordPress, along with YouTube, Vimeo, Soundcloud, all allow students to easily brand themselves in social media. Some of this technology was not available to students several years ago, but today faculty are aware of how software and new hardware make it easier for students to create their own social media presence. Faculty must push students to use this technology and require them to complete assignments that will be beneficial to students as they sell themselves to potential employers.

Some students recognize why personal social media branding is important. One student wrote, "it allows oneself to reach people worldwide with their talents," while another wrote, "it's important because you advertise yourself to show the world what you can do." Understanding the importance is critical; actually creating the perfect presence is far different.

An unprofessional online presence can be damaging to a student's potential career plans. As employers are increasingly looking at online footprints when making hiring decisions, it is crucial that students have guidance and mentoring on creating a respectable online brand. Data found reflect an attempt by students to put into writing their personal brand, yet, trending phrases such as "in progress," "developing," "I don't know," "laid back," and "it's not really professional" show a need for faculty to spend a great deal of time helping the student to know more about themselves and their potential. Using class time to instruct and work with students on personal brand development provided a safe environment where students could explore and produce an online image that reflected their authentic self and was professionally presented. Data from the open-ended questions showed students reporting themselves as "professional," yet there are few specifics given on what makes them professional. Assignments need to assist students in developing brands that are unique as opposed to generic. Students need to choose tools/platforms that highlight their skill sets. As the survey indicated, LinkedIn was an underutilized resource. Students should be encouraged to use this platform as a way to emphasize professional abilities, talents, and experience. The ability to create websites using simple-to-use templates and customized coloring and design provides students the opportunity to synthesize all of their experiences into one web location where a potential employer can readily access and then learn about the student's talents. In addition, students need to be reminded that personal brands are dynamic and need to evolve. They should be revisted and revised on a quarterly basis, along with a reexamination of personal long-term and short-term career goals. Brands that are allowed to stagnate and remain unrevised year after year can be almost as damaging as an unprofessional brand, by revealing a lack of discipline and focus.

REFERENCES

Arruda, W. (2013, October). Three steps for transforming employees into brand ambassadors. *Forbes*. Retrieved from http://www.forbes.com/sites/williamarruda/2013/10/08/three-steps-for-transforming-employees-into-brand-ambassadors/

Burnside, S. (2014, April 14). Investing in a brilliant new YOU™: The rise and tyranny of the "personal brand." *The Guardian*. Retrieved from http://www.theguardian.com/commentisfree/2014/apr/15/investing-in-a-brilliant-new-youtm-the-rise-and-tyranny-of-the-personal-brand

Dutta, S. (2010, November). What's your personal social media strategy? *Harvard Business Review, 88*(11), 127–130.

Edison, Arbitron, and Neilson. (2014). The infinite dial 2014. Retrieved from http://www.edisonresearch.com/home/archives/2014/03/the-infinite-dial-2014.php#.U4903HJdW8A

Gehl, R. (2011). Ladders, samurai, and blue collars: Personal branding in Web 2.0. *First Monday. 16*(9). Retrieved from http://firstmonday.org/ojs/index.php/fm/article/view/3579/3041

Goldsmith, M. (2009). Authentic personal branding. Retrieved from http://www.businessweek.com/managing/content/sep2009/ca20090929_228578.htm

Labrecque, L. I., Markos, E., & Milne. G. R. (2011). Online personal branding: Processes, challenges, and implications. *Journal of Interactive Marketing, 25*, 37–50.

Lair, D. J., Sullivan, K., & Cheney, G. (2005). Marketization and the recasting of the professional self: The rhetoric and ethics of personal branding. *Management Communication Quarterly, 18,* 307–343.

Langston, P. (2014, April). Hunting for work in the digital age. *Ottawa Citizen*. Retrieved from http://www.ottawacitizen.com/business/Hunting+work+digital/9724821/story.html

Marwick, A., & Boyd, D. (2010). I tweet honestly, I tweet passionately: Twitter users, context collapse, and the imagined audience. *New Media Society, 13*(1), 114–133.

Morgan, M. (2011, August). Personal branding: Create your value proposition. *Strategic Finan*ce, 13–14.

Peters, T. (1997). The brand called you. Retrieved from http://www.fastcompany.com/28905/brand-called-you

Smith, J. (2013, April). Why every job seeker should have a personal website, and what it should include. *Forbes.* Retrieved from http://www.forbes.com/sites/jacquelynsmith/2013/04/26/why-every-job-seeker-should-have-a-personal-website-and-what-it-should-include/

Umar, B. (2014, March). Five easy steps to discovering your personal brand. *The Huffington Post.* Retrieved from http://www.huffingtonpost.ca/bobby-umar/personal-brand_b_5051797.html

Wetsch, L. R. (2012). A personal branding assignment using social media. *Journal of Advertising Education. 16*(1), 30–36.

ASSIGNMENT 3.1. DEFINE YOUR PERSONAL BRAND

- I am an average college student, business/comical, clean and simple, creative, creative—innovative—diligent, developed but not strong—shows who I am and what I believe in—true representation of myself I believe.
- I'm a single female at Arkansas State University who is in a relationship, Delta Zeta sorority, and enjoy being involved on campus.
- I attempt to come across as genuine and honest (two very desirable personality traits) by being consistent and mostly positive.
- This personal brand is important because potential employers expect integrity in employees and friends and family have expectations of the same on a relationship level.
- It's not really professional.
- It's more of a person wanting to work in entertainment.
- My brand is about my professional broadcasting experience, my desire for continuing communications research, my work for my community, my love of God and my family.
- Traditional yet creative; my personal brand is unique to me.
- My personal brand would be one that shows my ability as a video maker and sports announcer.
- On social media, my image is centered around my family more than anything else.
- Reliant, always up for a challenge, innovative, helpful.
- Youthful and Expressive.

ASSIGNMENT 3.2. DISCOVERING YOUR PERSONAL BRAND

In the law, we talk about the discovery process. That's where the attorneys go around collecting all the information they are going to need for their case. It's the same thing with personal branding. I want you to spend some serious time reflecting on who you are and what you are about. It's not the girl in the sorority party pics or your obsession with Justin Beiber. Answer these questions to help you get to the core of who you are:

1. What would your fantasy/dream job be?
2. Describe yourself in three words.
3. What makes you unique? How are you unlike anyone else?
4. What skills give you the sense of timelessness?
5. What do you want your legacy to be?

Everyone take a few minutes to jot down your gut reactions to those questions. Write down what first comes to your mind. I'll give you four or five minutes.

Now, take what you have answered about yourself and distill it into about three sentences. This is a descriptor you can use on your social media sites that will be creative, consistent, and professional. But, don't stop there:

- Keep researching YOU; Define yourself based on what you find; Strategically select the right tools; Get wired; Maintain it; Experiment; and Connect

ASSIGNMENT 3.3. MEDIA MANAGEMENT AND ENTREPRENEURSHIP PROJECT

Phase 1

Assignment 1. Students are required to create a working résumé on the second week of school. This résumé allows the instructor and students to identify strengths and weakness of the student's personal brand. While students are welcome to create a résumé of any format, the instructor strongly encourages the students to follow a template that showcases the student's work and course highlights.

Assignment 2. Students are required to read the most recent copies of the Infinite Dial Reports by Edison and Triton. After discovery of how other people are using legacy and social media, students must examine their own personal media usage.

Assignment 3. Students must create a LinkedIn account and a Twitter account. These accounts must be shared with the instructor who will then be linked to the student. Students must create a one-sentence description of themselves that will run across all of their social media.

Assignment 4. Students must take a test to determine their own entrepreneurial characteristics. Students provide a two-paragraph discussion about what they believe are their strengths and weaknesses toward being an entrepreneur. Then they answer what their personal brand would be based on the test results.

Phase 2

Students meet entrepreneurs and business leaders in the community who introduce the concepts of strong personal credit, current market stability and instability, and tools needed to research a market such a demographics and psychographics.

Assignment 1. Students must do a SWOT analysis for their personal brand or their personal business.

Phase 3

Students are introduced in lecture situations to Reference USA, Small Business and Development tools, and other resources that are free for the student to use to learn about personal market. This market is identified as the place they want to be when they establish themselves.

Assignment 1. Students must complete a market analysis. Again there are numerous templates available for free that can be used to assist the student in doing a full analysis of the market. This is one of the most important assignments in this phase because the student must understand their market.

Phase 4

This is the final phase of the semester process of developing personal brands and product brands for the class. All information becomes a culmination of previously created documentation.

Assignment 1. Creating the business plan. The business plan uses data collected from the SWOT analysis and the Market Analysis to develop a coherent description and presentation of the brand. The executive summary is the most important two-page description of the student's brand; the writer introduces the public to the personal brand or product.

Assignment 2. Students resubmit revised résumé with additions to the content. This résumé will be used on the student's personal branding website.

Assignment 3. Students create a personal branding website.

Assignment 4. Final branding assignment. Students present a PowerPoint showcasing their personal brand or product.

CHAPTER FOUR

Social Media AND Applied Learning

HANA S. NOOR AL-DEEN

Social media have been growing at a remarkable rate while becoming widely available for facilitating various types of communication that are quickly delivered worldwide. They have been basically "the intersection of technology, social interaction, and information sharing" (Pavlik & McIntosh, 2015, p. 188). Social media have been considered "what used to be word of mouth and allowed it to now be heard by millions and spread virally with a nearly instantaneous reaction" (Thompson, Hertzberg, & Sullivan, 2013, p. 1). Due to such influx, numerous business organizations have been incorporating social media into their operational plans. For example, "among Fortune 500 companies, 73% now have company Twitter accounts and 66% have Facebook Pages" (Vella, 2012, p. 2).

The marketing and advertising industries have been among the early adopters of social media and they have integrated them into their business operations with the objectives of increasing exposure, building brand loyalty, developing a network of followers, maintaining business relationships with their customers, and so forth. Hettche and Clayton (2012) pointed out that "for advertisers and marketing practitioners, social media present myriad opportunities to engage customers, release product information and provide resources for an opt-in interactive environment" (p. 45). So, social media presence has become an essential component for numerous businesses. Thompson et al. (2013) maintained, "for most companies, opting out of the social media world is no longer feasible" (p. 1). Similarly, Florentine (2015) stated, "a social media presence for your business is no longer optional. It's a necessary part of your marketing, communications and customer relationship strategy in today's digital marketplace" (para.1). It is important to note that social media presence also demands a well-skilled staff to run such media. Florentine expounded "without a solid social media presence and professionals to manage that, your business will lack a major mouthpiece to reach potential customers, communicate effectively with your customer base and risk being considered out of touch or obsolete" (2015, p. 3).

Hence, businesses such as advertising and marketing have been primarily recruiting digital natives who are considered uniquely qualified to manage and operate social media for them. Digital natives are those who were born after 1990 and grew up during the era of the rise of digital technologies (DeGraff, 2014; Noren, 2011; Schroer, n.d.). It is interesting that a number of business organizations realized that some of this new staff was not quite ready to manage and operate social media at the business level even though such staff seemed to be skillful in using them at the personal level. Bennington (2015) observed that "just because someone is a 'digital native' does *not* make them uniquely qualified to manage your online branding" (p. 1). Yegulalp (2011) noted that

> social media mistakes . . . are bad enough in a personal context, although they can usually be straightened out. But when such things happen with a corporate Twitter account or some other branded outlet, they can be messier by orders of magnitude. (para. 3)

So, the need for skilled staff to manage and use the most effective strategies in operating social media has become vital for business organizations striving to establish a solid social media presence.

Consequently, many universities and colleges have been integrating social media into various areas of studies including advertising and marketing to address the needs of business organizations. Such a strategy is set to provide students with a learning opportunity for enhancing and channeling their personal use of social media into skills that are beneficial for businesses. Thus, this chapter is designed to offer students a learning approach that operates within the framework of the applied learning theory in order to empower them to transform their personal skills of social media into the advantageous use of them at a business level such as in advertising or marketing. This approach is carried out by providing students with a theoretical foundation along with practical application of using social media for advertising. The learning objective of this process is to provide students with hands-on experience in using social media to (a) promote their business and (b) maintain a business relationship with their customers. This course is principally designed for college students at the undergraduate level, but it can be easily adapted for a graduate-level course.

LITERATURE REVIEW

Although the adoption of social media by advertising and marketing business organizations has been on the rise, most of those businesses have not yet used social media to a maximum potential due to the need for social media specialists to manage and operate such media for them. Vella (2012) pointed out that "while businesses are hungry to tap social media, they lack the expertise to do so" (p. 2). He added, "among 2,100 companies surveyed by Harvard Business Review, a meager 12% of those using social media feel they use it effectively. The result is an exceptional demand for social media professionals who can boost the bottom line" (p. 2). The 2015 Social Media Marketing Industry Report revealed the results of the study that "surveyed over 3700 marketers with the goal of understanding how they're using social media to grow and promote their businesses" (Stelzner, 2015, p. 4). The findings showed that

> At least 87% of marketers surveyed felt they are struggling to answer all of the following questions: #1: Tactics: What social tactics are most effective? . . . #2: Engagement: What are the best ways to engage my audience with social media? . . . #3: Measurement: How do I measure the return on my social media marketing? . . . #4: Audience: How do I find my target audience with social media? . . . #5: Tools: What are the best social management tools? (Stelzner, 2015, p. 6)

DeMers (2014) added, "97% of marketers are currently participating in social media—but 85% of participants aren't sure what social media tools are the best to use" (p. 1). Therefore, business organizations have been faced with a challenge inasmuch as there has been a genuine need to use social media, yet there has also been a compelling need to have skilled staff to effectively manage and operate social media for them.

Education

A number of universities and colleges have incorporated social media into their curricula, albeit on a limited basis, due to the needs for specialists in social media even though the majority of them have been moving slowly to embrace such an important subject area. The irony is that "social networking may have been born in a dorm room. But when it comes to equipping students with the social media skills demanded by today's jobs, colleges are failing miserably" (Vella, 2012, p. 1). Dr. William Ward of Syracuse University's S.I. Newhouse School of Public Communications pointed out, "overall, the higher education system is failing to prepare students with the needed digital and social skill set in any meaningful way" (Vella, 2012, p. 1). Moreover, some in academia fail to understand the difference between "teaching social media" and "teaching the application of social media."

Teaching students "social media" would contribute little to the educational process of the students because the majority of them are already digital native/social media native, and quite frankly they often know more about social media than their instructors. However, teaching students "the application of social media" would contribute tremendously to the educational process of the students because they need to learn through hands-on experience how to transform their skills of social media into using them at the business level so that they can compete in the job market. Vella (2012) pointed out "while U.S. unemployment hovers around 8%, job postings requiring social media skills rose 87% from 2011 to 2012, topping 13,000 in one month alone earlier this year" (p. 2). So, this course afforded students a learning opportunity for hands-on experience within the framework of the applied learning theory.

Applied Learning

Applied learning has been generally categorized as experiential learning, hands-on experience, real-world experience, and so on while described in various ways. For example, applied learning has been defined as "an approach that contextualises learning in a way that empowers and motivates students, while assisting them to develop key skills and knowledge required for employment, further education and active participation in their communities" (VCAL, 2014, para. 1). As another example, "it designates the kinds of pedagogical principles and practices associated with engaged scholarship, communities of practice, civic engagement, experiential education, and critical pedagogy" (Schwartzman & Henry, 2009, p. 4). As a further example, "applied learning is to see it as *learning that is applied* to real world needs and challenges, which naturally require a combination of theory and application for individual, practical and academic growth in negotiated learning settings" (Renshaw, 2012, p. 2). Lester (2012) elaborated

> the success applied learning has . . . received its most notable endorsement from the Association to Advance Collegiate Schools of Business (AACSB). The recognition by AACSB of significant student learning taking place through personal involvement was a tribute to the educational archetype. (p. 119)

Such an approach to learning can be underscored as "'Activities that engage the learner directly in the phenomena being studied and are associated with structured reflection on the connection between the phenomena and theoretical concepts'" (Wolff & Tinney, 2006, p. 2). Ash and Clayton (2009) added, "each applied learning pedagogy provides students with opportunities to connect theory and practice, to learn in unfamiliar contexts, to interact with others unlike themselves, and to practice using knowledge and skills" (p. 25).

Accordingly, this social media course was planned to provide students with a learning approach that is operated within the framework of the applied learning theory while using a theoretical foundation and practical application as it affords the opportunity for the students to have hands-on experience to transform their personal skills of social media into the usage of such media at a business level for advertising or marketing. "The need for hands-on learning in advertising education has been expressed for decades" (Kinsky, 2015, p. 29). Lester (2012) put it simply as "theory and application come together in this applied social media project to yield an opulent learning environment where concepts and philosophies presented in the classroom are incorporated into strategies that culminate in unique creative productions" (pp. 119–120).

METHOD

Essentially, this course was intended for providing students with a theoretical foundation and practical application of using social media for advertising in a social media college course at the undergraduate level. In offering such a course, it would be important to take into account the student prior knowledge of using social media due to the fact that a digital divide has been prevalent for some time among those entering college in recent decades. In fact, it has been advocated that student prior knowledge of using social media should be evaluated because it is extremely significant for the educational process. Hettche and Clayton (2012) recommended a four-step process to measure student prior knowledge of social media by observing that

> While there are undoubtedly a number of ways that student prior knowledge of social media technology can be assessed and integrated into active learning exercises, we recommend a four-step process that involves (1) student prior knowledge assessment, (2) the integration of a specific course learning goal, (3) a synthesis between steps 1 and 2 and (4) the implementation of an active learning exercise. (p. 47)

Thus, student prior knowledge of using social media was assessed through an open-ended question survey during the first week of the semester of this course. Facebook and Twitter were the focus of the assessment because they were the primary social media used for advertising in this course. Research indicated that "Facebook (93%)" and "Twitter (79%)" were among "the top seven" social media that were used by marketers (Stelzner, 2015, p. 23). The results of the assessment revealed that student prior knowledge level was disproportionate with regards to these two social media platforms. Some of those who enrolled in the course had either used such social media sporadically at the personal level or they did not even have social media accounts for Facebook and/or Twitter. Subsequently, creating individual Facebook and Twitter accounts for those students took place immediately during the first week of the semester. Students who were active users of Facebook or Twitter were assigned to assist other classmates who were unfamiliar with using such social media in setting up their social media accounts while sharing with those classmates information that they deemed necessary about these social media accounts. The advantages of this approach were (a) creating a sense of community among classmates and (b) efficiency with regards to class time. Meanwhile, private Facebook Group and Twitter Group accounts were also created for the class. Two students were assigned to maintain a Facebook Group account and a Twitter Group account, respectively, so that they would gain experience of managing a social media site for a group at the class level. This task was rotated among those who were interested in obtaining such experience on a weekly basis.

At this point, the basic components for the course were established and the class was ready to move forward to reach its learning objective of providing students with hands-on experience in using social media for advertising to (a) promote their business and (b) maintain a business relationship with their customers. Operating within the framework of the applied learning theory, the learning process of this course was completed through *Theoretical Foundation* and *Practical Application*. The *Theoretical Foundation* was divided into two phases: *Theoretical Foundation (Phase One)* and *Theoretical Foundation (Phase Two)*. The *Practical Application*, meanwhile, was intended for presenting students with the opportunity of actual hands-on experience in using social media for advertising at a business level while maintaining business relationships with their customers. Through these assignments, students were also given the opportunity to experience working independently as well as working collaboratively with others through the paired assignment and the group project.

APPLICATIONS

I. Theoretical Foundation

A. Phase One (Weeks 1–4). Students were provided with the basic principles and generalizations of existing theories in relation to social media and advertising. Lectures, class discussions, Twitter discussions, case studies analysis, and critical analysis of specific topics related to both social media and advertising as well as social media and customers' business relationships were covered during this phase. Social media policies, codes of ethics, and best practices were also included due to the significance of this component of social media.

Among the topics that were covered during this phase for social media (i.e., Facebook and Twitter) and advertising were the articles "The Current Trends in Social Media Usage at Corporations: Analysis of Facebook Fan Pages of Fortune

500 Companies" (Shin, Carithers, Lee, Graham, & Hendricks, 2013, pp. 62–79) and "Twitter as Gateway to Relationship Marketing: A Content Analysis of Relationship Building via Twitter" (Watkins & Lewis, 2013, pp. 25–44). Similarly, social media and customers' business relationships were reviewed through case studies such as "Whole Foods Market: an extended community," and "Southwest Airlines: levels of engagement" (Long, 2012, pp. 150–154). During the class period, students were instructed to visit the social media sites of some of the organizations involved in these case studies and to observe those businesses' usage of social media and their customers' business relationship operations. Finally, social media policies along with current and previous court cases, codes of ethics, and best practices such as "Legal and Ethical Use of Social Media for Strategic Communicators" were discussed in detail (Stewart & Coleman, 2013, pp. 180–198). Some of the discussions took place face-to-face in the classroom while other discussions were carried out on the Twitter Group account of the class.

It is noteworthy that this course put equal emphasis on the learning process of both social media and advertising as well as social media and customers' business relationships. It has been repeatedly stressed that maintaining a positive business relationship with the customers is important for the marketing and advertising firms. Arens, Schaefer, and Weigold (2012) pointed out

> Most marketers know that the key to building brand equity in the twenty-first century is to develop interdependent, mutually satisfying relationships with customers. A market-driven firm's overriding purpose is to create happy, loyal customers. Customers, not products, are the lifeblood of the business. This realization has moved firms from simply *transactional marketing* to relationship marketing—creating, maintaining, and enhancing long-term relationships with customers and other stakeholders that result in exchanges of information and other things of mutual value. (p. 181)

Lawrence (2012) added, "it's far easier (about 50% easier according to Marketing Metrics) to sell to existing customers than to brand new prospects" (para. 1). Arens et al. (2012) explained that

> Defensive marketing typically costs less than offensive marketing because it is hard to lure satisfied customers away from competitors. The fragmentation of media audiences and the resistance of sophisticated customers to advertising messages make it increasingly difficult to succeed merely by stepping up the advertising volume. In fact, it costs five to eight times as much in marketing, advertising, and promotion to acquire a new customer as it does to keep an existing one. (p. 182)

B. Phase Two (Weeks 5–8). This phase gave students the opportunity to deal with two types of field research methods while focusing on social media, namely face-to-face interview and direct observations. The interview (i.e., *Local Business Interview and Social Media Usage*) was carried out face-to-face while the direct observations (i.e., *Online Business Analysis of Social Media Implementation*) were attained through *online direct observations.*

Phase Two—Part 1. Local Business Interview and Social Media Usage (Week 5). This assignment was conducted in one week and it was carried out prior to the *Online Business Analysis of Social Media Implementation assignment*. Each student in this class was required to conduct a face-to-face interview (using, to some degree, open-ended questions) with staff of a local business. The students were given the opportunity to select a local business based on their interest, but they had to be certain that the business already had a social media presence and that the social media have been actively used. Five interview questions were designed for this assignment, which included the two primary questions related to the strategies used to promote their business and to maintain business relationships with their customers (see Assignment 4.1). The students were also permitted to add any follow-up questions that they deemed necessary. Subsequently, the students were required to submit a typed reflective report that focused on the questions with the interviewee's answers as well as to give a three- to five-minute in-class presentation of their interview so that they could share their findings with the rest of the class members.

Phase Two—Part 2. Online Business Analysis of Social Media Implementation (Weeks 6–8). This assignment was given a three-week period for completion so that the students would have sufficient time to process this aspect of the course. Students were required to work in pairs for this assignment, and each pair of students handled one online business. But prior to giving the assignment to the students, a mini-workshop that was run by the instructor took place to prepare students for their online direct business observations while focusing on the usage of social media by such business organizations and how such business maintained business relationships with their customers. Thereafter, the students were given the assignment to conduct online direct business observations and to follow such business organization on its Facebook and Twitter accounts for two weeks (see Assignment 4.2). Students were given the option to select a for-profit or nonprofit business organization regardless of its size and type and then to submit an assignment proposal for approval prior to the observations. During that class period, students met in pairs to select their business for observation while working on their assignment proposal. The business organizations that were selected were local and national (both for-profit and nonprofit organizations). Regarding their observations, students were instructed to focus on the objective of the assignment, which was to ascertain the strategy that this business organization implements to promote their business and to maintain business relationships with their customers. Furthermore, the students were required to critically analyze whether or not such business organization has been using social media effectively with regards to their promotion and their business relationships with their customers. If the students concluded that their observed business was utilizing social media effectively, then the students had to explain how they reached that conclusion and to support it with examples. However, if the students concluded that the business organization was not using social media effectively, the students had to include recommendations for improving such an operation. Such critical analysis was based on the students' class training during the mini-workshop mentioned above as well as on the prior information that was obtained from the business interview assignment. Such preparation helped students to derive their own conclusions about the social media strategy used by the observed business.

Although three weeks were allocated for processing this assignment, the period of time used for the actual online observations was two consecutive weeks. Over that time, students were required to visit both social media of the selected business at least three times per week (on non-consecutive days), to spend at least 15 minutes per visit per student, to record the date and time of their visits, to keep a detailed journal of their observations during those visits, and to save each of the social media sites as PDF at the end of each visit as a reference. While the time spent observing social media was relatively short, it was sufficient for the students to grasp the essence of the business strategy of the observed business. Students were also encouraged to alternate the time and the day of the visits so that they could get a better overall picture about the business strategy that their observed business had implemented. The final step of this assignment consisted of the students submitting a typed reflective report while detailing their observations and critical analysis of their observed business, along with their social media journals and PDFs. Furthermore, the students were required to give a five- to seven-minute presentation before the class so that they would share their findings with their classmates.

II. Practical Application

Applying Social Media for Promotion (Weeks 9–13). Up to that point, students had gained adequate information through the theoretical foundation of this learning process with regards to the usage of social media for advertising as well as for maintaining business relationships with the customers. This last section was carried out through a group project that focused on empowering students to achieve their learning objective, which was gaining skills through hands-on experience about promoting a business on social media while maintaining business relationships with their customers.

At the beginning of Week 9, the students were given the project requirements (see Assignment 4.3). To implement this project, the class was divided into small groups consisting of three students per group. The students were required to *promote a hypothetical product, a service, or an event* (for a for-profit or a nonprofit business organization) *for three weeks* (Weeks 10–12) using three social media while maintaining a business relationship with their customers. The students were also given the option to promote a business that already existed provided that it did not have a presence on social media or perhaps it had a poor presence on social media. Facebook and Twitter were to be used in addition to a third social media platform of their choice. When selecting a third social media platform, they were advised to keep in mind the platform's importance for their business. The students were allowed to meet with their group members in the class during the first week of this project to brainstorm a selection of their intended business for the project as well as selecting a third social media platform and subsequently working on a project proposal for approval by the instructor. During that time, students were also able to secure instantly their instructor's advice regarding their business selections and any other aspect related to the project. Some students selected hypothetical businesses while others selected businesses that their families owned but that did not have a presence on social media or that had a rather poor presence on social media. The students' primary objective in promoting the businesses was either to create awareness or to increase awareness about such businesses while their secondary objective was to maintain business relationships with their customers. The students were also required to create three social media accounts (Facebook, Twitter, and a third social media platform) for their business. Most of the students selected Instagram in addition to Facebook and Twitter as their third social media platform. Interestingly, "Instagram (36%)" was ranked in the seventh place among "the top seven" social media that marketers used (Stelzner, 2015, p. 23).

Moreover, students were required to actively work on their social media site at least three times per week (non-consecutive days) during a period of three consecutive weeks, spend at least 15 minutes per visit, record the date and time of their visits, keep a detailed journal entry of their tasks during such visits, and save the social media site as PDF at the end of each visit as a reference. Thus, each student would basically spend a total of 2 hours and 15 minutes per week visiting the three social media sites; and, the individual social media site would be visited for a total of 2 hours and 15 minutes by the three students each week. It is worth mentioning that Stelzner (2015) reported "64% of marketers are using social media for 6 hours or more and 41% for 11 or more hours weekly. It's interesting to note that nearly 19% of marketers spend more than 20 hours each week on social media" (p. 14). The students were also encouraged to alternate the time and the day of their visits to the social media sites in order to gain a better perspective of their customers. The final steps of this project required the students to (a) submit the project package, which included the following components: business profile, social media strategy including SWOT analysis and promotional strategy, tracking report, reflective report, social media journals, and PDFs; and (b) make a five- to seven-minute presentation before the entire class so that they could share with their classmates the outcome of their project along with their personal hands-on experience of this learning process.

On completion of this project, the students had been presented with the opportunity of actual hands-on experience in using social media for advertising to promote a business while maintaining business relationships with their customers to achieve their learning objective. Their subsequent anonymous feedback demonstrated that this course had positively affected their learning process. For example, one student stated, "I gained so much knowledge about how to utilize social media from a business perspective," while another said, "Gaining first-hand experience was a huge plus of taking this class." An additional student mentioned that "there were a number of projects to complete in this class but they were enjoyable to complete."

Essentially, the students revealed an overall and genuine sense of confidence in explaining as well as demonstrating the strategy that they applied to (a) promote their business on social media and (b) maintain a business relationship with their customers. This indeed underscored that students had learned to transform their personal skills of social media into suitable ways for using such social media skills at a business level, thereby achieving the learning objective of this course. Notably, the students recognized that the beneficial take-away from this course was that they had become enabled to use social media at the business level coupled with a new and candid sense of self-assurance that they could compete in the social media advertising job market.

DISCUSSION

Social media have changed the way society communicates, learns, and conducts business. They have become powerful tools for marketers, advertisers, and many other business organizations (for-profit, nonprofit, and not-for-profit regardless of size and type) that, in turn, have elevated the need for skilled teams or individuals to manage and effectively operate social media for such organizations. Although the majority of digital natives are adept in using social media, there still exists a digital divide among this segment of society. Still, most digital natives tend to be skillful in using social media at the personal level, but such expertise might not be sufficient for application at a business level. While some businesses have been willing to invest the time and resources to train such employees, it remains a basic economic fact that the majority of them simply have not. Moreover, the majority of the businesses that use social media are small operations with limited resources at their disposal. The 2015 Social Media Marketing Industry Report showed that 60% of the businesses that participated in the survey ranged from self-employed to 10 employees (Stelzner, 2015). Still, almost two-thirds of the marketers plan to increase their use of social media in Twitter, YouTube, LinkedIn, and Facebook (Stelzner, 2015). Ultimately, the integration of social media into business organizations has brought with it opportunities as well as challenges. Among the challenges has been that of filling the demand for skilled college graduates in the area of social media for business organizations. Universities and colleges have accordingly approached fulfilling the business needs for a qualified social media workforce by incorporating social media into their curricula to prepare their students for this area of the job market. It was indicated that "students with social media certification are getting better jobs and internships," and "those who harness social communications are in high demand and have an advantage" (Vella, 2012, p. 2).

This chapter has demonstrated a teaching method for using social media for advertising in an undergraduate college level class through applied learning. It is important to note that throughout the three assignments, the learning objective intentionally remained the same for the purpose of reinforcement. Although this course took place at the undergraduate level, it can be easily adapted to a graduate-level course. It also showed that, by providing a theoretical foundation (such as face-to-face business interviews, online business observations, critical analysis, and case studies) and practical application (such as hands-on experience of using social media for promoting business plus collaborating with other classmates on achieving such a task), students were able to gain knowledge and develop skills germane to real-life simulations, which could prove quite helpful in transforming their personal skills of social media into skills that can be used for businesses such as in advertising and marketing. Along with such a process, there comes a greater sense of confidence on the part of the students when they seek to compete for social media advertising jobs in the employment market.

Finally, incorporating applied learning is essential for providing students with "experiential opportunities to promote student engagement in learning" (Sanders & Ming, 2014, p. 19). "Experience has long been considered the best teacher of knowledge" (Stephenson, 1998, para. 3). "This guided process helps students find meaning in their experiences and achieve growth in academic content, personal/professional development, and societal engagement" (Elmore & Murrill, 2014, p. 233). The convergence of social media with advertising and marketing has made providing students with practical application a must. Such application can empower students with the indispensable knowledge and skills so that they can successfully compete in the job market.

REFERENCES

Arens, W. F., Schaefer, D. H., & Weigold, M. F. (2012). *Advertising*. New York, NY: McGraw-Hill Irwin.

Ash, S. L., & Clayton, P. H. (2009). Generating, deepening, and documenting learning: The power of critical reflection in applied learning. *Journal of Applied Learning in Higher Education, 1*, 25–48. Retrieved from http://community.vcu.edu/media/community-engagement/pdfs/AshandClayton.pdf

Bennington, E. (2015). Social media strategy: Is it time to hire a social media officer? Retrieved from http://hiring.monster.com/hr/hr-best-practices/recruiting-hiring-advice/attracting-job-candidates/social-media-strategy.aspx

DeGraff, J. (2014). Digital natives vs. digital immigrants. Retrieved from http://www.huffingtonpost.com/jeff-degraff/digital-natives-vs-digita_b_5499606.html

DeMers, J. (2014). The top 10 benefits of social media marketing. *Forbes.* Retrieved from http://www.forbes.com/sites/jaysondemers/2014/08/11/the-top-10-benefits-of-social-media-marketing/

Elmore, K. S., & Murrill, L. D. (2014, February). *Outside the classroom walls: Transforming student perspectives through meaningful service-learning opportunities.* Paper presented at the Conference on Higher Education Pedagogy, Blacksburg, VA. Abstract retrieved from http://www.cider.vt.edu/conference/proceedings/2014ConferenceProceedings.pdf

Florentine, S. (2015). Should you hire a chief social media officer? Retrieved from http://www.cio.com/article/2865287/careers-staffing/should-you-hire-a-chief-social-media-officer.html

Hettche, M., & Clayton, M. J. (2012). Using social media to teach social media advertising: How to leverage student prior knowledge and WordPress blogs. *Journal of Advertising Education, 16*(1), 45–55.

Kinsky, E. S. (2015). This means war: Using an advertising war room simulation to teach social media skills. *Journal of Advertising Education, 19*(1), 29–41.

Lawrence, A. (2012). Five customer retention tips for entrepreneurs. *Forbes.* Retrieved from http://www.forbes.com/sites/alexlawrence/2012/11/01/five-customer-retention-tips-for-entrepreneurs/

Lester, D. H. (2012). Social media: Changing advertising education. *Online Journal of Communication and Media Technologies, 2*(1). Retrieved from http://www.ojcmt.net/articles/21/216.pdf

Long, M. C. (2012). Beyond the press release: Social media as a tool for consumer engagement. In H. S. Noor Al-Deen & J. A. Hendricks (Eds.), *Social media usage and impact* (pp. 15–154). Lanham, MD: Lexington Books.

Noren, L. (2011). Who is the millennial generation? *Pew Research.* Retrieved from http://thesocietypages.org/graphicsociology/2011/10/04/who-is-the-millennial-generation-pew-research

Pavlik, J. V., & McIntosh, S. (2015). *Converging Media* (4th ed.). New York, NY: Oxford University Press.

Renshaw, J. (2012). Learning twigs: Adventures in educational technology and design by learning. Retrieved from http://jasonrenshaw.typepad.com/jason_renshaws_web_log/2012/05/vcal-approaching-applied-learning.html

Sanders, C., & Ming, K. (2014, February). *A conversation about the integration of experiential learning opportunities in and outside of the classroom.* Paper presented at the Conference on Higher Education Pedagogy, Blacksburg, VA. Abstract retrieved from http://www.cider.vt.edu/conference/proceedings/2014ConferenceProceedings.pdf

Schroer, W. J. (n.d.). Generations X, Y, Z and the others. Retrieved from http://www.socialmarketing.org/newsletter/features/generation1.htm

Schwartzman, R., & Henry, K. B. (2009). From celebration to critical investigation: Charting the course of scholarship in applied learning. *Journal of Applied Learning in Higher Education, 1*, 3–23. Retrieved from https://www.missouriwestern.edu/appliedlearning/wp-content/uploads/sites/206/2013/09/Journal-of-Applied-Learning-in-Higher-Education-Volume-1-online.pdf

Shin, J-H., Carithers, H., Lee, S., Graham, M., & Hendricks, N. (2013). The current trends in social media usage at corporations: Analysis of Facebook fan pages of Fortune 500 companies. In H. S. Noor Al-Deen & J. A. Hendricks (Eds.), *Social media and strategic communications* (pp. 62–79). London, UK: Palgrave Macmillan.

Stelzner, M. A. (2015, May). 2015 social media marketing industry report: How marketers are using social media to grow their businesses. *Social Media Examiner.* Retrieved from http://www.socialmediaexaminer.com/SocialMediaMarketingIndustryReport2015.pdf

Stephenson, K. (1998). What knowledge tears apart, networks make whole. *Internal Communication Focus, 36.* Retrieved from http://www.netform.com/html/icf.pdf

Stewart, D., & Coleman, C. A. (2013). Legal and ethical use of social media for strategic communicators. In H. S. Noor Al-Deen & J. A. Hendricks (Eds.), *Social media and strategic communications* (pp. 180–198). London, UK: Palgrave Macmillan.

Thompson, T., Jr., Hertzberg, J., & Sullivan, M. (2013). Social media risks and rewards. *Grant Thornton.* Retrieved from https://www.grantthornton.com/~/media/content-page-files/advisory/pdfs/2013/ADV-social-media-survey.ashx

VCAL information sheet applied learning. (2014). *Victorian Curriculum and Assessment Authority.* Retrieved from http://www.vcaa.vic.edu.au/Documents/vcal/infosheets/VCALinfoapplearn.pdf

Vella, M. (2012, September). Universities are failing at teaching social media. *Fortune.* Retrieved from http://fortune.com/2012/09/26/universities-are-failing-at-teaching-social-media/

Watkins, B. A., & Lewis, R. (2013). Twitter as gateway to relationship marketing: A content analysis of relationship building via Twitter. In H. S. Noor Al-Deen & J. A. Hendricks (Eds.), *Social media and strategic communications* (pp. 25–44). London, UK: Palgrave Macmillan.

Wolff, M. K., & Tinney, S. M. (2006). Applied learning as a best practice model: A strategy for higher education student success. Baltimore, MD: University of Maryland.

Yegulalp, S. (2011). Social media disaster recovery: A first responder's guide. Retrieved from http://www.cio.com/article/2402497/social-media/social-media-disaster-recovery--a-first-responder-s-guide.html

ASSIGNMENT 4.1. LOCAL BUSINESS INTERVIEW AND SOCIAL MEDIA USAGE

Conduct an interview with a local business to ascertain the strategy that such businesses utilize regarding social media to (a) promote their business and (b) maintain the business relationships with their customers. Use the questions below as your guide for the interview to reach your objective above. You may also add any follow-up questions that you deem necessary. Subsequently, submit a typed reflective report that includes the questions along with the interviewee's respective answers. Each question needs to be addressed in a separate paragraph and your paper should consist of two single-spaced, typed pages.

Interview Questions and Answers

1. What is the business background? Include the link to your selected business along with the name and occupation of the person you interviewed as well as the date and time of the interview.
2. How do they reach their customers?
3. How has social media influenced the business?
4. What strategies have been used to promote this business through social media and what social media have they been using the most?
5. What strategies have been followed to maintain business relationships with its customers through social media?

Class Presentation

The presentation grade is prorated based on the quality of the content and of the presentation. Presentation Time: 3–5 minutes

ASSIGNMENT 4.2. ONLINE BUSINESS ANALYSIS OF SOCIAL MEDIA IMPLEMENTATION

Select a for-profit or nonprofit organization regardless of its size and type and make sure that such an organization has Twitter and Facebook accounts. Follow the organization on its Facebook and Twitter accounts for two consecutive weeks. Visit the social media of your selected business organization three times per week and spend at least 15 minutes per visit for your observations. Try to vary the times of your visits so that you can get a better overall picture about the business strategy. Keep a journal of your visits to its Facebook site and a journal of your visits to its Twitter site, record such entries in a bulleted format, and include the date and time of your visits. At the end of each visit, save each of the social media sites as PDF.

The objective of this assignment is to ascertain the strategy that this business organization implements to (a) promote their business on such social media and (b) maintain business relationships with their customers though the usage of these social media.

The following questions need to be addressed: (1) What is the business background? (2) Based on your observations, do you think that this business organization has been utilizing social media effectively with regards to their promotion and their business relationships with their customers? (3) If you concluded that your observed business has been utilizing social media effectively, explain how and provide examples to support your conclusion. However, if you concluded that they were not using social media effectively, what would you recommend for improving such an operation?

Assignment Proposal (50–100 words)

Submit a typed assignment proposal (no later than this class period) to get the approval from the instructor for your work. The proposal should include the name of the organization along with its links and why you selected such an organization.

Package Submission

The following components need to be submitted by the day of your presentation of this assignment: Reflective Report (300 words); Facebook Journal; Facebook PDF; Twitter Journal; and Twitter PDF.

Class Presentation

The presentation grade is prorated based on the quality of the content and of the presentation. Presentation Time: 5–7 minutes (each individual).

ASSIGNMENT 4.3. APPLYING SOCIAL MEDIA FOR PROMOTION PROJECT

Promote a hypothetical product, a service, or an event (for-profit or nonprofit business organization) for three consecutive weeks using three social media. If you wish to promote a business that already exists but it does not have a presence on social media or it has a poor presence on social media, you may do so. Facebook and Twitter must be used for such promotion in addition to a third social media platform of your choice. When selecting a third social media platform, keep in mind its importance for your selected business. Through this project, you will gain hands-on experience in using your personal skills of social media at a professional level. *Note: In promoting your hypothetical business, you need to include a disclaimer to that effect on your social media sites accordingly.*

The objective of this project is to (a) demonstrate the promotional strategy that you apply to promote such business and (b) explain the strategy that you utilize to maintain business relationships with your customers. You need to visit your social media sites at least three times per week, spend at least 15 minutes per visit per site, record the date and time of your visits, keep a detailed journal entry of your tasks during such visits, and save the social media sites as PDF at the end of each visit. Try to alternate the times of your visits to better understand your target market.

Project Proposal

Submit a typed project proposal including the title/name of the business, a brief description of it, and the three social media platforms that you plan to use for it *no later than the next class period.* You also need to include why you selected this business. I encourage you to evaluate thoroughly your project proposal before submitting it.

Package Submission

1. Business Profile
   Describe (a) the business background, (b) your promotional objective(s), and (c) the target market.
2. Social Media Strategy
   a. Create three social media accounts (Facebook, Twitter, and one additional social media platform for your business).
   b. Elaborate on the creative strategy and the forms that you employed for your promotion on social media.
   c. Discuss SWOT analysis of your business while remaining mindful of your social media.
   d. Explain your promotional strategy that you applied to promote your business on social media as well as the strategy that you used to maintain business relationships with your customers.
3. Tracking Report
   Submit a social media tracking report showing the fluctuations of your business traffic during the promotional period.

4. **Additional Items**
   The following components (including the artifacts) need to be submitted electronically inside their respective folders:
   Reflective Report (300 words); Facebook Journal; Facebook PDF; Twitter Journal; Twitter PDF; the Journal of the third social media platform; and the PDF of the third social media platform

Class Presentation

The presentation grade is prorated based on the quality of the content and of the presentation. Presentation time: 5–7 minutes (each individual).

PART TWO

Applying Social Media IN Teaching Public Relations

CHAPTER FIVE

Taming THE Social Media Data Deluge

Using Social Media Research Methods in the Public Relations Classroom

KATHLEEN STANSBERRY

The use of online technologies has been growing at an incredible rate. In 2012 the online marketing and research company DOMO published a wildly popular infographic showing the volume of new media content created and shared every minute (James, 2012). Using data sourced from the most popular online social networking platforms, DOMO found that each minute Facebook users shared nearly 700,000 pieces of content, Twitter users sent more than 100,000 tweets, YouTube users uploaded 48 hours of video, and Instagram users published more than 3,500 new photos. In 2014 DOMO updated their popular infographic and found that in just two years' time the rate of online content generation had exploded. Information released in April 2014 showed that Facebook activity increased to nearly 2.5 million per minute, the number of tweets almost tripled to 277,000 per minute, YouTube uploads increased to 72 hours of new video per minute, and Instagram postings jumped to 216,000 new photos published each minute (James, 2014).

The growth in shared content was largely fueled by the increasing popularity of social media among Internet users. According to a 2014 survey conducted by the Pew Research Center, approximately 52% of the adults in the United States who are active online participated in two or more social media sites (Duggan, Ellison, Lampe, Lenhart, & Madden, 2014). This represented a significant increase from 2013, when Pew data showed only 42% of Internet users active on two or more social networks (Duggan & Smith, 2013). While Facebook was by far the most popular social network identified in the Pew study, more than a quarter of online adults reported using programs such as Instagram and Pinterest, which facilitate the production and exchange of visual content (Duggan et al., 2014). The percentage of individuals who used social media to share multimedia content has risen rapidly, and it has become imperative that future public relations professionals be equipped with the skills to research and measure this popular form of communication.

Social media users have been leaving digital footprints that showed their activities, interests, ideas, connections, and concerns. The process of making sense out of this cacophony of social media content, however, could be intimidating for even experienced strategic communications professionals. The public relations industry has changed dramatically as social media use has become increasingly prevalent. Practitioners have overall reported a belief that social media have enhanced public relations practice, particularly in regards to improving interactions with external audiences (Wright & Hinson, 2013). Approximately 69% of the Fortune 2000 companies used social networking sites as part of their strategic communication efforts (McCorkindale, 2010). The interactive nature of social media platforms make ideal spaces to engage in dialog with members of key publics, and numerous studies (e.g., Rybalko & Seltzer, 2010; Smith, 2013; Sommerfeldt, Kent, & Taylor, 2012) have examined the dialogic potential of social media platforms. However, the use of social media within the practice of public relations has gone far beyond the conversational properties of social networks. The information shared by key publics on social media sites has been a gold mine for public relations practitioners looking to understand the concerns, needs, and preferences of their target audiences. The purpose of this chapter is to present several active-learning assignments that can be used in public relations research courses to give students experience working with social media data and applying social media research in strategic communication campaigns.

LITERATURE REVIEW

Audience Research

The 2014 Communication and Public Relations Generally Accepted Practices (GAP) report produced by USC Annenberg Strategic Communication and Public Relations Center (SCPRC) found that organizations were increasingly using audience research in their campaign planning process (Swerling et al., 2014). According to the report, about a third of the public relations practitioners surveyed indicated that they measured online conversations and more than 70% strongly agreed that there was a need for research-oriented public relations professionals who could interpret data and apply it in practice.

Recent research in audience behavior has focused primarily, although not exclusively, on the concept of participation (Livingstone, 2013). The shift in audience studies has developed in conjunction with the rapid rise in population of digital technologies and the act of participation was inexorably connected to the realities of our increasingly mediated lifestyle (Hepp, 2013). McQuail (2013) pointed out that the players in the message dissemination process could no longer be neatly categorized into senders (mass media) and receivers (the audience). Audiences received content, but they also produced, transformed, categorized, and redistributed messages through social media platforms. Audiences actively participated in not just the consumption of media but the production and distribution of mediated messages as well.

More than half of adult Internet users reported posting and sharing original multimedia content such as photos or videos (Duggan, 2013). Near-ubiquitous cell phone ownership, coupled with the rising popularity of photo and video sharing apps such as Vine, Instagram, and Snapchat, indicated that the number of audience members who also acted as content creators and distributors would continue to grow (Duggan, 2013). The unprecedented power of audiences to participate in the media production and distribution process has prompted a new wave of interest in publics research in the field of public relations (Vardeman-Winter & Tindall, 2011).

Audience as Publics

The study and practice of public relations has traditionally segmented audiences into "publics," or groups of people concerned with a specific issue (Rawlins & Bowen, 2005). The situational theory of publics, first proposed by Grunig (1968) and updated and refined over the next several decades (Grunig, 1978, 1983, 1987, 1989, 1992, 1997; Grunig, Grunig, & Toth, 2007; Grunig & Hunt, 1984), set the foundation for much of the study of publics in the field of public relations and was a mainstay of public relations textbooks. The situational theory of publics described publics as coming into being in response to issues that affect the members of that public. The theory recognized four different types of publics based on their response to issues: non-publics, latent publics, aware publics, and active publics. Groups were categorized according to three independent variables: problem recognition, constrain recognition, and level of involvement (Grunig & Hunt, 1984).

Public relations professionals tended to be most concerned with members of active publics (groups with high problem recognition, low constraint recognition, and high levels of involvement) because they were likely to mobilize in response to a particular issue and would take action that could affect organizations (Grunig, 1978, 1983, 1997; Grunig & Hunt, 1984). The more active a public, the more members of that public participated in shared communication behavior, and the more likely individuals were to develop shared attitudes and participate in similar behaviors in relation to their common issue (Grunig & Ipes, 1983). Publics that communicated actively about an issue were likely to develop solutions to shared problems and strategies to address shared concerns (Grunig & Stamm, 1979).

Social media platforms provided a space for active and aware publics to engage in communication activities centered on shared concerns. Due to the public nature of social media platforms, these conversations were often available to public relations researchers and practitioners that had the skill set to collect and analyze social media data.

Public Relations Research Pedagogy

Research was one of the essential elements of the four-step strategic planning process widely used in public relations—research, planning, implementation, and evaluation (Broom, 2009). The Public Relations Society of America (PRSA), the leading professional organization in the field of public relations, recommended that public relations students at both the undergraduate and graduate levels receive a thorough grounding in research methods and analysis. According to a 1999 report by the Commission on Public Relations Education (CPRE), the ideal undergraduate major in public relations would include a course in "public relations research, measurement, and evaluation" (p. 3). In a follow-up report that looked specifically at public relations education at the master's level (CPRE, 2012), the CPRE recommended that qualitative and quantitative research methods, including historical-critical, focus groups, case studies, interviewing, sampling, surveys, and experimental, be included as part of all public relations master's degree programs.

While traditional public relations research methods were a staple of modern public relations education, practitioners were increasingly using social media to engage with publics (Sha, 2011) and the next generation of practitioners would need to expand their toolsets to take full advantage of the research opportunities that online, participatory communication platforms afford. Practitioners reported a particular need for students to understand a variety of digital research and analytics tools to take advantage of the different insights available through social media (Anderson & Swenson, 2013). The following sections describe a five-week learning module that would provide students with experience in the practice of advanced social media research and analytics techniques.

METHOD

The overarching philosophy that informed the development of the applications described in this piece was one of experiential learning. Practitioners and academics alike supported a hands-on approach to teaching major concepts in public relations (Aldoory & Wrigley, 2000). Case studies and the use of real-world clients were mainstays of public relations pedagogy and were well-suited to the learning style of millennial students who tended to prefer experiential learning (Gallicano, 2011). The five-week learning module described in the following sections provided numerous opportunities for students to explore social media tools for the purpose of performing research on behalf of a client.

Research is an integral part of most public relations capstone courses, a time when students are likely to work with clients on campaign planning and development projects. However, much of the research in these courses focused on client, rather than audience, research (Benigni, I-Huei, & Cameron, 2004). The following learning activity incorporates the use of experiential learning through the use of real-world clients or case studies and introduces students to several new media research techniques that can be used to better understand the interests and concerns of publics as they relate to a particular topic or organization. This section describes a five-week learning module that walked students through the process of researching the opinions and concerns of sample clients' key publics by accessing information that was shared via social media (see Assignment 5.1). Each week students worked to master a particular social media research technique by engaging in hands-on exercises in which they gathered information about their client produced by active social media users. This module could also be easily expanded to a 10- or even 15-week learning experience depending on the amount of time spent on classroom instruction. These assignments covered the basic processes of applying network analysis, content analysis, influencer identification, and data-mining techniques when studying social media networks. The learning module cumulated with a final paper that required students to produce a report that analyzed audience perceptions of a clients' online brand based on their research findings.

This project would be most appropriate for upper-level public relations classes. It would be particularly well-suited for classes that focus on public relations campaigns, strategic social media use, and promotional communication research techniques.

The learning module described below was first completed by 18 students in a capstone class in public relations. Students were divided into groups of three and assigned a client based on their interests and declared preferences. The instructor worked with a university service learning coordinator to identify appropriate local clients that would benefit from the research activities performed by the advanced public relations students. The clients represented a broad range of local nonprofit institutions and included a cancer resource organization, veterans, support and rehabilitation group, historical and fine arts museum, and women's roller derby club.

APPLICATIONS

Social Media Research Project: Five-Week Learning Module

Week 1: Choosing a Client and Identifying Key Publics

The social media research project can be used with real-world clients in a classroom setting or can be applied in a case study situation. When used as part of a case study project, it is recommended that the instructor provide students with a detailed client description and situation analysis that clearly identifies the goals, concerns, and communication needs of the sample client. There are several online resources available for instructors looking for cases and sample clients. For example, the Public Relations Society of America (PRSA) maintains an extensive database of case studies that describe Silver Anvil Award–winning public relations campaigns. In addition, examples from the annual Public Relations Student Society of America (PRSSA) Bateman Case Study Competition are available at no charge on the PRSSA website.

Although working with real clients might require additional time spent preparing for classes and more complicated logistical processes for setting up client meetings and managing client-student interactions, both students and faculty members reported that client projects offer enhanced learning opportunities for public relations students (Aldoory & Wrigley, 2000). While students could work one-on-one with a client, this project was designed to be completed in small groups. The recommended group size for each client team was anywhere from three to five students. Although more than one group could research on behalf of the same client, no student or student team should be responsible for more than one client over the course of this learning activity.

Once clients were identified, students developed a list of key publics based on the client's situation and goals. If, for example, a local coffee shop with the goal of expanding their customer base had acted as client, key publics may have included coffee aficionados, area college students, and other independent small business owners. For each key public, students identified distinguishing demographic, psychographic, and technographic characteristics.

While the use of demographic data and psychographic data have long been standard methods of market segmentation, technographic segmentation has been a relatively new concept. The technographic profile, an increasingly common method of market segmentation, focused on the "motivations, usage patterns, and attitudes about a technology" (Potter, Forrest, Sapolsky, & Ware, 1988, p. 36). Forrester, a global research and professional advisory firm, offers several free online tools to assist in consumer technographics research. Understanding a public's role in social media content creation and distribution would be important in later aspects of this learning project, and the technographics profile should be completed as thoroughly as possible.

Assignment. The assignment for the first week of this learning module (a) required students to create profiles for their client's key publics that incorporated demographic, psychographic, and technographic information; (b) introduced students to the Forrester Social Technographics Consumer Profile tool; and (c) required students to think critically about their client's situation and goals to identify key publics. Students were asked to create a detailed description of two key publics that they would target as part of a communication campaign on behalf of their client. As this was a capstone course, the students had previous experience identifying and describing target audiences as part of the public relations planning process. Each target public description included demographic, psychographic, and technographic profile information.

To determine technographic profile data, students were required to use Forrester's Social Technographics Profile tool. The free online tool enabled users to input basic demographic information of age, country of residence, and gender. Based on these characteristics, the tool then categorized the resulting audience segment by their likely use of social media. The technographics profile showed the likely percentage of creators, social media users who created and shared original content; critics, users who commented on blog posts or left reviews on shared websites or mobile applications; collectors, users who categorized and collated online content; joiners, those who actively maintained a profile on social networking platforms such as Facebook or LinkedIn; spectators, social media lurkers who participated primarily by viewing content that others post; and inactives, or Internet users who did not participate in social media (Li, 2007).

Students drafted and turned in profile pieces on two different key publics. In addition to including appropriate details regarding demographic, psychographic, and technographic features, the profiles also clearly stated the relationship between their client and the identified publics.

Week 2: Social Media Audit and Content Analysis

During the second week of this learning module students conducted a social media audit and topic-based content analysis to determine how clients were represented online and what they could learn from ongoing social media conversations relating to their brand and/or organizational goals. As the class met twice a week, the first class session was devoted to lecture on the purpose and best practices when performing a social media audit and online content analysis, and the second session was left open for students to work on this project within their groups. Students also worked on the assignment outside of class time.

Social Media Audit. The social media audit is a subset of the common public relations research tool the communication audit. According to Elsasser (1994), a communication audit did three things: "1) Reviews how well an organization communicates internally and externally. 2) Focuses on messages exchanged, media used, and the outputs and outcomes of the communications process. 3) Helps identify missing linkages or blocked circuits in the communications process" (p. 1). The social media audit is similar to a traditional communication audit, but instead of looking at the entirely of an organization's internal and external communication messages, the audit only addresses the social media messages produced and distributed by an organization.

For the purposes of this assignment, students performed a social media audit for their client. The students reviewed clients' web pages, Facebook profiles, Twitter accounts, LinkedIn pages, Wikipedia entries, and other social media platforms as appropriate. Based on their findings, students analyzed their client's social media presence and developed a SWOT analysis, identifying the organization's strengths, weaknesses, opportunities, and threats as related to their online presence. The SWOT analysis format was chosen as a way to format the social media audit both to give students experience using this method of organization and because it provided a succinct structure for client reporting purposes.

Content Analysis. The second part of the assignment for Week 2 of the learning module required students to complete a basic qualitative and quantitative content analysis based on social media content relevant to the key publics identified in Week 1. Hootsuite, a popular social media management and tracking tool, was used to collect social media mentions for analysis. Each group opened a free Hootsuite account for the purposes of this assignment. They then identified at least three search terms that would be relevant to both their client and their client's key publics. For example, the student group that worked with a local cancer outreach and support organization used the following search terms: *cancer in Ohio*, *cancer survivor*, and *pink ribbon*. The instructor recommended students use the free search insights tool Google Trends to help identify popular search terms within their various subject areas. These search terms were entered into Hootsuite's search tool, creating a real-time database of social media conversations. Students who were unfamiliar with the Hootsuite system were directed to the extensive list of guides and tutorials available through Hootsuite's online Help Desk.

Because all of the students in the class had passed a required research methods course prior to being accepted into the public relations major, they were familiar with the concept of qualitative and quantitative data analysis. To complete the Week 2 assignment, each group produced a report describing what they learned from the collected social media messages. Sample quantitative data included number of mentions, use of multimedia content, relative popularity, and sentiment (i.e., number of positive mentions vs. negative mentions). For the qualitative analysis, students identified major themes that emerged from content of the collected messages. For example, the student group that worked with the roller derby team found that "player support," or messages directed at particular roller derby participants by fans, was a major theme of social media mentions related to the alternative sport.

Given that students had only a week to complete both a social media audit and a qualitative and quantitative content analysis, the data collected and analyzed represented a fairly basic understanding of the multifaceted conversations occurring online around any given topic. However, this exercise did provide students with a general understanding of how to perform a social media audit, identify search terms, and analyze social media messages.

Week 3: Identifying Influencers Using Online Social Network Analysis

The third week of the learning module introduced students to the process of online social network analysis. Social network analysis was used to show the web of connections that helped define a public. In this online social network analysis assignment, students collected and mapped the hyperlinks between webpages to provide visual representations of at least one of the key publics identified in Week 1 of the learning module. By mapping links between websites, students were able to identify influencers in online, interest-based publics and provide insight into the communication flow process throughout the network.

Online Social Network Analysis. In social network analysis, the actors within a network were called nodes and the connections between those actors were called ties, links, or connections. Social network analysis has long been used to map social connections among individuals where the ties or connections refer to interdependencies such as friendship, sexual history, financial relationships, and common interests (Hanneman & Riddle, 2005). In online social network analysis, network-mapping software was used to create visualizations of the connections between web presences. The analysis of automated network crawls, which used a starting set of webpages to launch a crawl that iteratively follow hyperlinks to map the interconnected nature of online sites, has added a new dimension to the study of online communication practices in recent years. The process used by network crawlers was similar, although implemented on a much smaller scale, to the process of large search engines such as Google or Yahoo, which used automated crawlers to search and collate content throughout the web. Automated network crawls produced social network data that positioned websites as individual actors (or nodes) within a network and hyperlinks as the connection between these actors.

IssueCrawler. To complete this assignment, student teams used the free online web-mapping tool IssueCrawler. IssueCrawler has been a publicly available crawler system offered by the Amsterdam-based Govcom Foundation. The processing power for the tool was shared among the registered researchers using it, much like early mainframe computers shared processing power at the advent of the Internet. IssueCrawler recorded and followed the hyperlinks from a starting point indicated by the user. Each student group created a list of at least 10 bloggers that they believed were influential within their clients' key publics. These blogs acted as the starting point, or seed nodes, from which the network crawl was launched.

The software can be programmed to perform both co-link analyses and snowball analyses. When performing a snowball network crawl, IssueCrawler identified and recorded all links present in the seed nodes, then all outgoing links present in the websites linked to by the seed nodes. The number of times that this process took place was dictated by the depth of the crawl. In a co-link analysis, the crawler identified and recorded only sites that were linked by at least two of the starting points. This filtered out isolated sites and helped to ensure that crawls at deeper depths maintained ties to the original network. Because students used the network analysis tool to look for influential websites within their key publics, the teams ran co-link analyses with a crawl depth of two. IssueCrawler read, recorded, and followed all links on a webpage, including in-text links, sidebar links such as blogrolls, header links, and links to personal social media presences on Twitter and Facebook.

The visualizations presented by the students in this assignment all used the concept of degree centrality to determine strength of ties. Degree centrality is defined as the number of links connected to a node. In the case of directed networks such as these, where websites posted directional links to other sites, degree centrality is typically defined in two separate measures: indegree and outdegree. Indegree count referred to the number of links coming from other nodes and connecting to the site in question. Outdegree was the number of links that a network player directed to others. Because the students who completed this assignment were primarily concerned with the relative influence of websites within their key publics, they were instructed to use indegree centrality, which was interpreted as a form of popularity, as the primary measure to determine influence within the network visualizations. Websites on the network graphs created by IssueCrawler that were most tightly clustered with one another exhibited the highest measure of indegree centrality, and information that flew through these network players was most likely to be distributed across the full network.

Students were required to draft and submit a report on the results of the online network analysis each group produced on behalf of their client. The reports included images of the network maps created using IssueCrawler and an analysis of the network visualization, including a list of websites deemed highly influential within the community based on number of inbound links and indegree centrality. Many students found that the network mapping activity helped them identify the social media platforms that were particularly popular with members of their key publics. For example, a student group that looked at a network map created using a list of cancer blogs as seed nodes found that Twitter was one of the most influential sites within the network. By comparison, a network created using a list of museum websites as seed nodes found that Facebook and Flickr were highly influential within the resulting map.

Week 4: Content Tracking via Social Media

The public relations industry has long been concerned with the process of message dissemination. Early public relations practice focused almost exclusively on distributing information to various news media sources on behalf of clients. Media relations continued to be a primary function of public relations and more than 90% of public relations practitioners reported that media relations was one of their core responsibilities (Swerling et al., 2014). Although traditional media played a primary role in the process of message distribution, social media users have also been influential distributors of content.

This assignment introduced students to research techniques examining social media content distribution. This was particularly valuable when looking to understand what messages were frequently shared among an organization's key publics. Recent research has found that certain characteristics, such as content that evokes high-arousal emotions (Berger & Milkman, 2012), were more likely to be shared than others. However, it has also been found that the context within which a user views online content effected their likelihood of engaging in sharing behavior (Phelps, Lewis, Mobilio, Perry, & Raman, 2004). For the purposes of campaign planning, it is imperative that public relations practitioners understand what will resonate within their key publics.

BuzzSumo. In this assignment students used the social media search engine BuzzSumo to look at what content was most shared within a certain subject area. BuzzSumo is a tool that crawls through the links that people share on popular social networking sites such as Twitter, Facebook, LinkedIn, Google+, and Pinterest and ranks videos, photos, articles, and other online content according to number of shares. For example, the student group that worked with a veterans support organization searched for the term "Homeless Veterans," which returned a list of the most shared content related to the issue of homelessness among veterans. BuzzSumo allowed users to sort content by social network, type of content (article, infographic, video, interview, blog post, and so on), and timeframe (past day, week, month, six months). Users could also save their searches, which allowed them to return to the site throughout the campaign development process and made it easier to produce client reports.

In addition to searching for shared content, BuzzSumo featured an option to a search for influencers based on subject-specific search criteria. The resulting list of influencers could be filtered by type, including bloggers, companies, journalists, and "regular people." This search tool could be particularly valuable for public relations practitioners engaging in social media outreach and relationship building campaigns with influential members of key publics.

Students were required to submit an analysis of the findings of their BuzzSumo search process, including a description of the type of content that appeared to be most popular among members of the key publics identified in Week 1 of this project. Each team of students was instructed to perform and analyze at least two subject searches for each of their key publics and showed how the results might be beneficial to their clients.

Week 5: Final Analysis and Client Research Report

During the fifth and final week of the social media research learning module students used the situational theory of publics to structure a final analysis of the research project. In addition, each student team prepared a report for their client that provided a summary of their research findings.

Final Analysis. The final analysis assignment consisted of a short paper (five to seven pages) in which students were required to explain the situational theory of publics and show how it informed their research process. In the first part of the final analysis paper, students categorized the key publics identified in Week 1 of the learning module as latent, aware, or active, using examples from their research to show how the publics' online communication activity aligned around shared issues. Furthermore, students discussed what shared cognitions the key publics appeared to be forming around common issues.

In the second part of the final analysis paper, students analyzed how their research could benefit their client. Students were specifically asked to show how their clients could use information on primary influencers within key publics, knowledge of publics' common shared cognitions, and an understanding of frequently shared content could contribute to the development of successful persuasive communication campaigns. Students had the option of producing individual final analysis papers or working together on the paper with other members of their client team.

Client Research Report. The last piece of the social media learning module was the client research report. Students produced a written report summarizing their research findings and presented it to their clients. The report consisted of the following sections: Research Summary and Primary Findings, Description of Key Publics, Social Media Audit Findings, Content Analysis Findings, Online Social Network Analysis Findings, Content Tracking Findings, and Recommendations for Campaign Development. Each student team worked together to produce a single client report, which was then submitted to the instructor in hard copy. Once approved by the instructor, the client reports were submitted both electronically and in hard copy to the client contacts.

Evaluation. Project evaluations were completed at the end of a five-week term. Students reported feeling as though they learned research skills that would help them stand out as candidates in a crowded job market and claimed they appreciated the experiential nature of the research project. One common critique of the social media learning module was that students said they would have appreciated more time to master the new software tools and gain a more nuanced understanding of the research techniques. In general, students said the connection of the assignment to public relations theory, particularly the situational theory of publics, was clear and provided context for the work. Feedback from class clients, which was collected during informal discussions at the close of term, was also positive with all clients stating that the students had found useful information through the course of their research. Many clients indicated an interest in receiving training in the methods of social media research that the students had completed on their behalf.

DISCUSSION

The true value of social media to public relations practitioners is not as a channel to distribute messages to key publics, but as a way to learn about the concerns and needs of those publics. The social media research learning module described in this chapter exposed students to current research trends and tools that academics and practitioners alike have noted as valuable in current public relations practice. Although the five-week process allowed for only an introduction to various forms of social media research, student and client feedback indicated that even this limited amount of time spent exploring new research tools was useful within an experiential learning situation.

With the advent of participatory media, tools that enable web-based mass communication are now widely available, and technology-savvy influencers are able to spread information to other web users through a variety of online media. The increasing popularity of social media platforms has erased the already fine line between media producer and audience. Social media users can create and distribute original content and also play a significant role in extending the reach of traditional media messages, giving publics an unprecedented amount of influence. This changing media landscape makes it imperative that public relations practitioners devote research time to analyzing the interests and concerns of their key publics, particularly those active online. The overwhelmingly positive feedback provided by students who completed the social media research indicated that they understand the value of learning research methods that take advantage of social media content.

As mass media sources become increasingly fragmented, public relations professionals struggle to communicate with key publics (Kent & Taylor, 2002). Depending on the target audience, a viral video could be just as effective at increasing awareness among a public as an article in the *New York Times* or a lead story on the national nightly news. This fragmentation has led to many public relations practitioners discovering (or rediscovering) the importance of interpersonal communication in public relations (Kent & Taylor, 1998). As interpersonal communication practices are increasingly occurring in hypermediated environments, the public relations profession must adapt with revised strategies and tactics.

Instead of focusing the bulk of their efforts on communicating with key publics through mainstream media sources, public relations professionals should shift their efforts toward research and incorporation of the values and concerns of key publics. At the moment, public relations professionals largely use media relations tactics such as news releases, media advisories, and media events to communicate with active members of online publics. But online communities form around shared issues and concerns, not a desire to broadcast news and information to a broad audience. Members of online communities use media tools to facilitate interpersonal communications publically, but they are not members of the media and should not be treated as such. Online, interest-based publics consist of individuals with a shared passion for a common issue or interest. Teaching students to perform publics research not only exposes them to advanced social media analytics tools and techniques, it helps prepare them to thrive in a rapidly changing profession.

REFERENCES

Aldoory, L., & Wrigley, B. (2000). Exploring the use of real clients in the PR campaigns course. *Journalism & Mass Communication Educator, 54*(4), 47–58.

Anderson, B., & Swenson, R. (2013). What should we be teaching our students about digital PR? Collaborating with top industry bloggers and PR Twitter chat professional. *Teaching Public Relations, 87*, 1–4.

Benigni, V., I-Huei, C., & Cameron, G. T. (2004). The role of clients in the public relations campaigns course. *Journalism & Mass Communication Educator, 59*(3), 259–277.

Berger, J., & Milkman, K. L. (2012). What makes online content viral? *Journal of Marketing Research, 49*, 192–205.

Broom, G. M. (2009). *Cutlip & Center's effective public relations* (10th ed.). Englewood Cliffs, NJ: Prentice Hall.

Commission on Public Relations Education. (1999). Public relations for the 21st century: A port of entry. Retrieved from http://www.commpred.org/aportofentry/

Commission on Public Relations Education. (2012). Standards for a master's degree in public relations: Education for complexity. Retrieved from http://www.commpred.org/educatingforcomplexity/

Duggan, M. (2013). Photo and video sharing grow online. *Pew Research Center*. Retrieved from http://www.pewinternet.org/2013/10/28/photo-and-video-sharing-grow-online/

Duggan, M., Ellison, N., Lampe, C., Lenhart, A., & Madden, M. (2014). Social media update 2014. Pew Research Center. Retrieved from http://www.pewinternet.org/2015/01/09/social-media-update-2014/

Duggan, M., & Smith, A. (2013). Social media update. *Pew Research Center.* Retrieved from http://pewinternet.org/Reports/2013/Social-Media-Update.aspx

Elsasser, J. (1994). Doing a communications audit. *Public Relations Tactics. 12.*

Gallicano, T. D. (2011). Relationship management with the millennial generation of public relations agency employees. Paper presented at the meeting of the Association for Education in Journalism and Mass Communication. St. Louis, MO.

Grunig, J. E. (1968). Information, entrepreneurship, and economic development: A study of the decision-making process of Colombian Latifundistas. Unpublished doctoral dissertation, University of Wisconsin, Madison.

Grunig, J. E. (1978). Organizations and publics: Testing a communication theory. *Journalism Monographs, 46.*

Grunig, J. E. (1983). Communicating behaviors and attitudes of environmental publics: Two studies. *Journalism Monographs, 81.*

Grunig, J. E. (1987, May). When active publics become activists: Extending a situational theory of publics. Paper presented to the International Communication Association, Montreal, Canada.

Grunig, J. E. (1989). Sierra Club study shows who become activists. *Public Relations Review 15*(3), 3–24.

Grunig, J. E. (1992). *Excellence in public relations and communication management.* Hillsdale, NJ: Lawrence Erlbaum.

Grunig, J. E. (1997). A situational theory of publics: Conceptual history, recent challenges and new research. In D. Moss, T. MacManus, & D. Vercic (Eds.), *Public relations research: An international perspective* (pp. 3–48). London: International Thomson Business Press.

Grunig, J. E., Grunig, L. A., & Toth, E. L. (2007). *The future of excellence in public relations and communication management: Challenges for the next generation.* Mahwah, NJ: Erlbaum.

Grunig, J. E., & Hunt, T. (1984). *Managing public relations.* New York, NY: Holt, Rinehart & Winston.

Grunig, J. E., & Ipes, D. (1983). The anatomy of a campaign against drunk driving. *Public Relations Review, 9*(2), 36–52.

Grunig, J. E., & Stamm, K. R. (1979). Cognitive strategies for the resolution of environmental issues: A second study. *Journalism Quarterly, 56*(4), 713–720.

Hanneman, R. A., & Riddle, M. (2005). *Introduction to social network methods.* Riverside, CA: University of California, Riverside. Retrieved from http://faculty.ucr.edu/~hanneman/

Hepp, A. (2013). *Cultures of mediatization.* Cambridge, England: Polity Press.

James, J. (2012, June 8). Data never sleeps: How much data is created every minute? [Infographic]. Retrieved from http://www.domo.com/blog/2012/06/how-much-data-is-created-every-minute/

James, J. (2014, April 23). Data never sleeps 2.0: How much data is created every minute? [Infographic]. Retrieved from http://www.domo.com/blog/2014/04/data-never-sleeps-2-0/

Kent, M. L., & Taylor, M. (1998). Building dialogic relationships through the World Wide Web. *Public Relations Review, 24*(3), 321.

Kent, M. L., & Taylor, M. (2002). Toward a dialogic theory of public relations. *Public Relations Review, 28*(1), 21–37.

Li, C. (2007, April 23). Forrester's new social technographics report [Blog post]. Retrieved from http://forrester.typepad.com/groundswell/2007/04/forresters_new_.html

Livingstone, S. (2013). The participation paradigm in audience research. *Communication Review, 16*(1/2), 21–30. doi:10.1080/10714421.2013.757174

McCorkindale, T.C. (2010). Can you see the writing on my wall? A content analysis of the Fortune 50's Facebook social networking sites. *Public Relations Journal, (4)*3. Retrieved from http://www.prsa.org/Intelligence/PRJournal/Documents/2010McCorkindale.pdf

McQuail, D. (2013). The media audience: A brief biography—Stages of growth or paradigm change? *Communication Review, 16*(1/2), 9–20. doi:10.1080/10714421.2013. 757170

Phelps, J. E., Lewis, R., Mobilio, L., Perry, D., & Raman, N. (2004). Viral marketing or electronic word-of-mouth advertising: Examining consumer responses and motivations to pass along email. *Journal of Advertising Research, 45*(4), 333–348.

Potter, W., Forrest, E., Sapolsky, B. S., & Ware, W. (1988). Segmenting VCR owners. *Journal of Advertising Research, 28*(2), 29.

Rawlins, B., & Bowen, S. (2005). Publics. In R. L. Heath (Ed.), *Encyclopedia of public relations* (p. 718). Thousand Oaks, CA: SAGE.

Rybalko, S., & Seltzer, T. (2010). Dialogic communication in 140 characters or less: How Fortune 500 companies engage stakeholders using Twitter. *Public Relations Review, 36*(4), 336–341.

Sha, B. (2011). 2010 practice analysis: Professional competencies and work categories in public relations today. *Public Relations Review, 37*(3), 187–196.

Smith, B. (2013). Digital conversion: Social media, engagement, and the "I am a Mormon" campaign. *PRism (9)*1. Retrieved from http://www.prismjournal.org/ homepage.html

Sommerfeldt, E. J., Kent, M. L., & Taylor, M. (2012). Activist practitioner perspectives of website public relations: Why aren't activist websites fulfilling the dialogic promise? *Public Relations Review, 38*(2), 303–312.

Swerling, J., Thorson, K., Tenderich, B., Yang, A., Lis, Z., Gee, E., & Savastano, E. (2014). GAPVIII: Eighth communication and public relations generally accepted practices study. *USC Annenberg Strategic Communication and Public Relations Center*. Retrieved from http://annenberg.usc.edu/ResearchCenters/Strategic%20Communication%20and%20Public%20Relations%20Center/PrevGAP.aspx

Vardeman-Winter, J., & Tindall, N. T. J. (2011). The current climate on publics segmentation research: Publics at a nexus of multiple identities and digitisation. *PRism 8*(2). Retrieved from http://www.prismjournal.org/fileadmin/8_2/Vardeman-Winter_Tindall.pdf

Wright, D. K., & Hinson, M. D. (2013). An updated examination of social and emerging media use in public relations practice: A longitudinal analysis between 2006 and 2013. *Public Relations Journal*, 7(3), 1–39.

ASSIGNMENT 5.1. SOCIAL MEDIA RESEARCH PROJECT (FIVE-WEEK LEARNING MODULE)

Week 1: Key Publics (15%)

In this assignment you will identify and describe the client that you will research in this course. In addition, you will identify two appropriate key publics for your class client and create profiles that incorporate demographic, psychographic, and technographic information.

Assignment Guidelines:

- Your paper should include the following:
    - Description of class client
    - Client Goal
    - Description of two key publics. Each description must include:
        - Demographic Data
        - Psychographic Data
        - Technographic Data
            - Please use the Forrester Technographics Profile tool to develop a technographic profile for your key publics
              Link: http://empowered.forrester.com/tool_consumer.html

Week 2: Social Media Audit and Content Analysis (15%)

In this assignment you will conduct a social media audit and topic-based content analysis to determine how clients are currently represented on social media.

Assignment Guidelines:

- Your paper should include the following:
    - Social Media Audit
        - Provide a review and analysis of the following: client webpage, Facebook profile, Twitter account, LinkedIn page, Wikipedia entry, and other social media platforms as appropriate
    - SWOT Analysis
        - Identify client strengths, weaknesses, opportunities and threats
    - Content Analysis
        - Use HootSuite to monitor at least three search terms related to your client's goals and objectives and provide a written report of your findings.
          Link: https://hootsuite.com/

Week 3: Identifying Influencers (15%)

In this assignment you will collect and map the hyperlinks between webpages to provide visual representations of one of the key publics identified in the key publics assignment. In addition, you will analyze the network maps created in this assignment to identify influencers within your client's key public.

Assignment Guidelines:

- Use the online web mapping program IssueCrawler to create a network map representing one of the key publics identified in Assignment 1.
  Link: https://www.issuecrawler.net/
- Prepare and submit the following:
  - List of seed nodes (starting points for the web crawl)
  - Screenshot of network map representing the interest-based key public
  - Analysis of network map identifying primary influencers within the key public

Week 4: Content Tracking (15%)

In this assignment you will explore media relations in a social media context. To complete this assignment you will utilize the content mining tool BuzzSumo, which is a web-based program that ranks videos, photos, articles, and other online content according to number of shares.

Assignment Guidelines:

- Please use the content search tool BuzzSumo to explore two subject areas for each of the key publics identified in Assignment 1.
  Link: http://www.buzzsumo.com/
- Your paper should include the following:
  - Brief description of each of your two key publics
  - Subject-specific search results for each of your two key publics
  - Analysis of subject searchers including information on how the search results could be beneficial to clients in the campaign development process

Week 5: Final Analysis (20%) and Client Research Report (20%)

For this assignment you will complete both a final analysis of your research findings and a research report appropriate for submission to your client. Your work for these assignments represents the culmination of the social media research project.

Assignment Guidelines:

- Final Analysis (five to seven pages)
  - This paper requires you to use Grunig's situational theory of publics to structure a final analysis of the research you have completed during this learning experience. In this paper please answer the following questions:
    - Are the key publics you have researched through this project latent, active, or aware? Please explain your rationale for this categorization.
    - Explain how the research you have completed on primary influencers within key publics, knowledge of publics' common shared cognitions, and an understanding of frequently shared content could contribute to the development of successful persuasive communication campaigns.
- Client Research Report
  - You will complete a final report showing the results of your client research. This report should clearly represent the work you have completed on behalf of your client. The quality of your work should be appropriate for client submission, which means entirely free of grammatical errors and typos.

CHAPTER SIX

Drop AND Give Me 20 (Social Media Platforms)

Using Boot Camp to Teach Social Media Strategy

MELISSA L. JANOSKE, ROWENA L. BRIONES, AND JULIA DAISY FRAUSTINO

Teaching public relations at the undergraduate level now requires instructors to be well versed in social media. Fifty-six percent of corporate executives say that engaging with customers online is at least a top-10 priority for their organization (Brown, Sikes, & Willmott, 2013), and 75% of companies have a dedicated social media team (Solis & Li, 2013). In recent years, demand for public relations practitioners with social media knowledge and expertise skyrocketed (Lombardi, 2013), as more employers pursued practitioners with more specialized social media skills involving platforms such as Instagram, Vine, and Twitter (Stone, 2013). As Wright and Hinson (2010) discovered in their longitudinal survey, 85% of practitioners claimed that social media platforms have made a lasting impact on their organizations by changing the strategy behind their communication. Thus, preparing students to handle social media strategically and effectively, as well as helping them understand the variety of platforms that exist to discern which are most relevant to their organization, is now a crucial skill for helping them succeed in the public relations job market.

While students often seem to believe that their personal use of social media easily translates into a professional or business understanding of social media, oftentimes this might not be the case. In many instances, public relations students need guidance on how they can translate personal social media uses into more academic and professional understandings and contexts (Melton & Hicks, 2011). Seventy-one percent of executives believed there were risks involved in using social media for their organization, but they also believed that those risks could be mitigated with education and strong knowledge of how different platforms can best be understood and implemented strategically (Thompson, Hertzberg, & Sullivan, 2013). In particular, more hands-on, experiential-learning focused assignments and projects could produce beneficial outcomes such as improving understanding of strategy, collaboration, technical skills, and networking (Fraustino, Briones, & Janoske, 2015; Melton & Hicks, 2011). Additionally, studies have shown that using social media in the classroom can create stronger interpersonal relationships between instructors and students, leading to a more positive classroom experience (Waters & Bortree, 2011). Such education and knowledge are the focus of the assignment for junior- and senior-level public relations students discussed in this chapter.

In discussing a social media "boot camp" assignment, this chapter looks to build understanding of the importance of social and cultural contexts in implementing social media strategy in public relations and underscore the need for models and demonstrations in effective teaching that facilitated a cyclical learning process for students. It also builds on knowledge of how to teach strategic use of social media for public relations purposes, culminating in insights and best practices gleaned from three authors that can assist future instructors with student engagement, experiential learning, and strategic thinking, along with the potential downsides of using social media in public relations assignments. The assignment also provided students with similar knowledge bases of social media platforms, including platform functions, personal and professional uses, costs, and distinctive abilities to aid in public relations work, which included those same downsides that students might otherwise choose to ignore due to their personal vested interests in social media. Through analysis of the work of these students, this chapter also offers a broader discussion of social media's place in public relations work.

LITERATURE REVIEW

Moody and Bates (2013) contended that university public relations programs face both opportunities and challenges in attempting to continuously adapt to and integrate industry trends in undergraduate curriculum. Because technology has been constantly evolving, instructors have been tasked with ensuring that their materials should be up-to-date and accurately reflect what current public relations practitioners face in the workplace. There remained a lack of uniformity in public relations curricula (Alexander, 2004; Briones & Toth, 2013), meaning public relations graduates emerged from different programs with an inconsistent continuum of knowledge and skills as they entered the job market.

At the onset of public relations curriculum development, universities were expected to focus on broad areas such as research, ethics, writing, strategic thinking, and theory (Commission on Public Relations Education, 2006). More recently, however, employers have emphasized the need for graduates to have skills and expertise in social media and technology in particular. This is evidenced by current trends in the industry; for instance, a recent report by Wanted Analytics showed that public relations specialists were the second most sought after professionals when it came to social media knowledge and expertise (Lombardi, 2013).

A challenge facing public relations instructors, however, has been overcoming students' misperceptions that they have already been social media experts because they were digital natives who grew up with these technologies. Public relations courses in social media needed to emphasize the difference between professional vs. personal use of these platforms, and how they could translate their knowledge into a strategic organizational context (Melton & Hicks, 2011). Industry professionals have already recognized this discrepancy; one study found that lack of new media skills was a major complaint coming from employers looking to hire public relations graduates (Todd, 2009). Therefore, constantly updating curricula by integrating skill and strategy development related to new media platforms and technologies has not only been beneficial for public relations students but also somewhat imperative if they would hope to stand a chance in an always shifting work environment. Furthermore, bridging academia and practice in the public relations classroom via social media has been serving to strengthen the instructor-student relationship as well; studies have demonstrated that encouraging social media engagement in the classroom humanized instructors and deepened student engagement (Fraustino et al., 2015; Waters & Bortree, 2011).

Experiential Learning

Perhaps particularly relevant to public relations as both a theory-driven and skill-based content area, scholars in many fields have turned to the theoretical lens of experiential learning to help craft courses that bridge theoretical and applied components. Experiential learning theory (ELT) is founded in the work of Kolb (1984), who conceived of learning as a process of cyclically integrating thoughts and experiences. ELT proponents asserted that developing knowledge required the "grasping" and "transforming" of experience (Kolb, 1984, p. 41). Grasping experience is achieved through having *concrete experience* and engaging in *abstract conceptualization* about it (or apprehension and comprehension, respectively), and transforming experience is achieved through enacting *reflective observation* and related *active experimentation* (or intension and extension, respectively) (Baker, Jensen, & Kolb, 2002; Kolb, 1984, p. 41). That is, students create knowledge through recurrently experiencing, reflecting, observing, and acting.

ELT has advanced several key ideas about what learning is and how it comes about (Kolb, 1984; Kolb & Kolb, 2005, 2009). For example, ELT contends that learning is based in processes rather than their outcomes. Learning includes and often requires relearning—and thus requires adaptation. Further, ELT supports the idea that learning can be prompted through the process of resolving conflicts between varying viewpoints and orientations, privileging reflection on differences. Moreover, learning does not exist in an individual bubble; the learner and the environment are inextricably interdependent, and actions between them facilitate learning. Ultimately, learning is the construction and reconstruction of knowledge. Learners engage these ideas through the cyclical progression of grasping and transforming experience.

The ELT conception of the learning process has been meaningful and promising for the public relations classroom. Indeed, public relations scholars, instructors, and students' potential employers have begun to endorse the value of experiential learning techniques in preparing public relations and communication students for their careers (Benecke & Bezuidenhout, 2011; Todd, 2009; Wilson, 2012). This approach has particularly been relevant in the arena of new and social media. Creating assignments and activities for students that encourage them to learn and relearn through experiencing and reexperiencing social media platforms, reflect strategically on those experiences, and expand on those reflections by engaging in subsequent adaptive experiences using similar and different platforms could tap into ELT and facilitate learning and creation of knowledge.

Connectivism

Connectivism has at its core the belief that "learning principles and processes should be reflective of underlying social environments" (Siemens, 2005, p. 1). Namely, this means that student learning should incorporate and reorganize itself to take advantage of technological advances and their impact on life and learning. This also takes into account the "half-life of knowledge," or the time from when something was learned until the information has become obsolete (Gonzalez, 2004, p. 1). As technology improves and information is shared more rapidly, the amount of time it takes to learn something decreases as the amount of time before that information has become obsolete increases. To balance the half-life of knowledge, instructors and organizations should look to new ways of gathering and understanding information. Connectivism, then, encourages technological experimentation and innovation, looking for engagement and creativity with the technology itself to improve learning (Ebben & Murphy, 2014).

Connectivism allows for learning to occur outside of an individual, by gathering groupings of information and forming connections between those groups and the individuals who understand them. When new information is continually being acquired, learning has become a process of gathering a diversity of opinions and connecting information sources together to make the most sense of the information presented (Siemens, 2005). Information flow is also important, as how the knowledge moves from one person or organization to another impacts whether an individual stays current through those connections (Siemens, 2005).

The Internet offers a unique opportunity for information flow and network connections to exist by allowing the large efforts of a few individuals to enhance the connections and efforts of the many through improving their own learning and knowledge (Brown, 2002). This network extension also includes the idea that knowing where to find information has become more important than knowing the information itself, leading to less emphasis on memorization; if people know where to find it, and can access it quickly and easily through technology, individuals are less likely to gather and store that information for themselves (Fitzsimmons, 2008). Based on this perspective, teaching now has become a way to model and demonstrate information, and learning has become a means for individuals to practice and reflect (Downes, 2007). Social media have been "inherently provocative," offering us a new way of thinking about how to gather and collaborate on information (Bonzo, 2010, p. 914). With the prevalence of digital information, and the changing ways students have been using technology to learn, the ability to process digital messages easily and clearly has become a skill students should be practicing within the classroom (Locker & Kienzler, 2012).

Thus, the above discussions of social media in the public relations classroom, experiential learning, and connectivism prompted the following research questions:

RQ1: What is the impact of the boot camp experience on student understanding of social media and public relations?

RQ2: How, if at all, does the boot camp assignment facilitate or detract from experiential learning and/or connectivism in relation to social media and public relations knowledge creation?

RQ3: What can be learned about best practices in teaching social media in a public relations classroom?

METHOD

A total of 101 students at three unique mid-Atlantic public university campuses were taught by three individual instructors. Three universities were chosen to broaden the perspective of how students would accept and construct the assignment, and to increase the number of platforms and examples available for discussion. The instructors were able to combine their understanding and experience to guide the creation and application of the assignment for the upper-level social media strategy classes, an assignment that took on the form of a boot camp (see Assignment 6.1).

The assignment was introduced at the beginning of the semester to institute base knowledge of a variety of social media platforms for the students. Allowing the students to choose the platform and example they were interested in increased student interest and engagement as well. The three instructors conducted classroom observations in their own classes during the presentations and throughout the semester. Students in all of the classes wrote memos about the experience. Two classes' student memos were content analyzed, for a total of 63 memos in the data set (one class' memos were lost due to an online technical failure). Qualitative content analysis took a grounded theory approach, using a constant comparative method to identify and categorize themes (Corbin & Strauss, 2008). Coding was first done in a line-by-line manner and then merged into categories through axial coding (Corbin & Strauss, 2008). These categories were primarily based on the three research questions of student understanding of social media and public relations, experiential learning and/or connectivism, and teaching best practices. The researchers discussed the coding to ensure consistency and to adapt codes and categories as necessary. Themes and categories were also discussed in relation to the reviewed literature, all of which allowed for the potential identification and removal of researcher bias (Corbin & Strauss, 2008). These themes are further explicated in detail in the application section below but were listed as follows: relationship to personal experience, thinking strategically about social media tools for particular clients, recognizing benefits and drawbacks, finding the right public(s), grasping and transforming experience, building a knowledge base through technology and connections, covering as many platforms as possible, fostering continued reflection, knowing the platforms, creating opportunities for real-world application, and discussing downsides.

New Media Boot Camp Assignment Description

Jovially named after the U.S. military new-recruit indoctrination colloquially referred to as "boot camp," the new media boot camp assignment was designed to quickly and efficiently help students know of and understand a wide variety of social media platforms, how they worked, and how or why they might be best utilized by public relations professionals. The assignment offered a list of 45 social media platforms suggested by the instructors, but students could also suggest other platforms for instructor approval. Although some of the most popular platforms such as Twitter, Facebook, Pinterest, and Instagram were included on the list, instructors encouraged students to choose a platform they were less familiar with as a way of broadening their understanding of social media. No topic could be presented twice in a single class. Because there were in some cases fewer students in the course than platforms on the list, some of the instructors either offered the option of a second presentation as extra credit or discussed remaining platforms during lectures and activities throughout the semester.

The assignment required each student to perform two sets of actions: (1) create a five- to seven-minute presentation about the platform, and (2) write a reflective memo about the boot camp experience. Presentations occurred in back-to-back class periods so students were immersed in the variety of platforms and could more easily recall and apply points of comparison and contrast. Memos were due one to two weeks after presentations concluded to allow time for thoughtful reflection on all platforms. The activity took place at the beginning of the semester to provide immediate foundational knowledge. Not everything about the platform could be covered in five to seven minutes, but the goal was to give students a working, top-level, practical understanding of the platform, creating a class-wide baseline knowledge that could be built upon thereafter.

The reflective memo asked students to discuss which platforms from boot camp they thought would be most helpful or effective for public relations professionals and why, as well as how the client they were working for that semester could utilize the platform (or, if they were not working for a client that semester, they could choose an organization and explain the potential benefit of that platform for the chosen organization). Memos were two pages, double spaced.

Students were graded on the presentation based on respecting the time limit and giving sufficient discussion of each of the four areas discussed above: background, usage, signing up, and strengths and weaknesses. Reflective memos were graded based on the overall discussion and the recommendations made for the client/chosen organization. Other aspects of the grade included creativity, grammar/spelling/punctuation/style, and appropriate citation of sources.

APPLICATIONS

RQ1: Student Understanding of Social Media and Public Relations

The first research question looked to uncover the ways boot camp allowed students to broaden their understanding of how social media can be applied in public relations work. Students used this assignment to talk about platforms that were both familiar and unfamiliar, and they found ways to integrate all of them into the work they did over the course of the semester as well as the work they envisioned themselves doing in the future. Students were also able to understand and see examples of a broader variety of platforms because three universities were involved. Key themes here included using their previous personal experience, thinking strategically about social media use for particular clients/organizations, recognizing benefits and drawbacks, and learning to find and understand a public.

Relationship to Personal Experience. Many students discussed the platforms in relation to their own personal experience with social media. One student discussed realizing the differences in personal and professional social media use, noting that some tools not enticing to a particular individual might still be smart tools for a business to devote time and resources. Another student commented on how the assignment increased awareness of unknown platforms, which she claimed is important for a young public relations practitioner. Students recognized that they were adding to their existing knowledge base.

Learning about the history and background of various platforms allowed students to see more about what the platforms offer, even for platforms students thought they had prior knowledge of—and they noted they could use such information to make decisions for clients. But when they did not have any experience with a particular platform, students also noted that and discussed how their expanded knowledge would benefit them. Students made it clear they expected that social media skills would help them on the job market and in public relations roles.

Thinking Strategically About Social Media Tools for Particular Clients. Students were also specific about the multiple ways social media could be used, either by the client they were working with for that semester or for a company or organization they hoped to work for in the future. Broadly, one student noted that "finding the right tool and learning how to use it correctly could make or break a public relations campaign effort." More specifically, another student talked about using what she learned from boot camp at her internship, explaining that using Klout could help her convince skeptical people in upper management that social media strategy would be beneficial for the organization. One student also discussed the idea that multiple platforms could be useful, and that a top priority for a public relations practitioner is to help the client.

Further, participating in boot camp allowed students to envision social media in a different sense. They noted that "many of the new platforms that were introduced during the boot camp are finding a place in day-to-day public relations tactics." This also allowed students to realize that "social media platforms are always changing and adapting . . . it is important that public relations professionals should think creatively in terms of social media." Finally, the students were able to recognize that the presentations during boot camp had a clear message: "don't underestimate what your organization can accomplish with social media."

Recognizing Benefits and Drawbacks. Students spent time in the memos looking to explain why or how they might utilize a particular platform, or why they might find it less useful for a specific client. They talked about platforms that were "cost friendly and easy to make," or "a great way for the public to connect with your brand." Platforms were lauded for being an "interesting twist" or "a smart move."

Sometimes, platforms were deemed useful not only in helping the organization but also in helping the public relations professional. One student suggested that because "part of the job deals with pitching presentations to clients . . . this tool is a crucial and informative tactic." Yet, despite identifying many pros of social media use for PR activities, students struggled to identify related issues. One of the few problems noted by students focused on using less well known platforms, such as "how many of these platforms will exist in the next year or so."

Finding the Right Public(s). Students demonstrated awareness that one major benefit of diversifying social media platform usage is the ability to converse with a variety of individuals. These platforms could either help them (or an organization) reach the proper public(s) and/or to understand more about who might make up that particular public from a demographic or psychographic standpoint. While some discussions of this idea were generally at a surface level, other students showed great depth and reflection about ways to learn about a public via social media. One student discussed trying to help the client, a local organization for dog lovers in the area, noting that "one of the questions we aim to answer through our research is, who is the typical dog owner? Through Twitter, demographics could be sought out to answer this question."

Many students also paid attention to the idea that not every platform presented at boot camp would be best for every organization, or for every public. Specifically, one noted that, "of course, depending on your age group you are targeting this might not be as important but it would still be smart to have a basic profile." Boot camp reflections prompted students to begin thinking about how extensively an organization should devote resources into using a platform specifically based on use (or not) by their target public(s). Another student talked about how "while these platforms are very interesting, their reach seems to be too large for the client."

RQ2: Experiential Learning and/or Connectivism

Student memos and instructor observations showed that the boot camp assignments facilitated experiential learning among students and also fostered conditions for connectivism. Key themes relevant to this research question included grasping and transforming experience as well as building a knowledge base through technology and connections.

Grasping and Transforming Experience. As noted above, the experiential learning framework conceived of learning as a process of creating knowledge through experiencing, reflecting, observing, and acting in a recurrent cycle. Boot camp allowed students to grasp the experience through researching and using various social media platforms and engaging others about the platforms in class through interactive demonstrations and then transformed that experience through reflective memos and application of learned tools to solve real-world public relations problems.

Indeed, all the key ideas underlying ELT were evident in student memos, comments in class, and instructor observations. For example, ELT asserted learning is construction and reconstruction of knowledge. Students exhibited this idea when creating presentations and reflecting on their and their classmates' work, and they noted it in their reflections. For example, a student explained the process of breaking down existing notions of obtained knowledge and rebuilding it, explaining that when entering the class he thought he had a solid knowledge of social media, but the assignment revealed to him that the breadth and depth of the new media landscape was far beyond his prior knowledge. Not only did students engage in abstract conceptualization but they also reflected on practical applications, rounding out an initial cycle of experiential learning. Several students devised specific strategic plans for use of particular platforms for existing or future clients and/or their current internships, putting new knowledge into practice. More broadly, one student concluded that "my classmates gave presentations that I walked away from with knowledge on a topic that I found to be necessary for utilization of strategic public relations." As the semester continued, students built and rebuilt ideas and experiences based on their boot camp activities.

Building a Knowledge Base Through Technology and Connections. Students were able to engage the notion of connectivism by seeking information both inside and outside of the classroom setting. First, by turning to digital resources and online communities to expand their knowledge base, students could learn in a manner quite different from the more traditional methods instructors typically use. This resulted in increasing the students' awareness levels on how influential technology could be. Second, the students used the network of their peers to learn more about social media tools that they perhaps had never heard of or used. Finally, students were excited about how researching the various platforms caused them to realize the full power and potential of the social media and how it can be useful in day-to-day public relations work. One student claimed, "after our social media boot camp I had a new perspective of the power of social media." They also discussed the ability of social media to build community and connections between various groups of people, one remarking that "networking is a large part of public relations." A student discussed online community building through another platform, Meetup.com, but talked about it in a different context: building client relations in a more professional setting.

RQ3: Teaching Best Practices

This research question looks at understanding what best practices for teaching social media may be gained from this study. Suggestions and perceptions from all three instructors were brought together here without identifying information, as they reached consensus on the ideas. Five best practices were identified, including (1) cover as many platforms as possible, (2) foster continued reflection, (3) know the platforms, (4) create opportunities for real-world application, and (5) discuss downsides.

Cover as Many Platforms as Possible. Students frequently mentioned the value of being exposed to multiple platforms, including those they thought they already knew and others with which they were unfamiliar. One major benefit of the boot camp structure was that it allowed everyone in the course to be on the same page about what specific platforms were, what they could do, and ways they could best be used in public relations. This expansive shared knowledge prompted richer discussions throughout the rest of the semester, especially in helping students continue to broaden their brainstorming ideas beyond default Facebook, Twitter, and/or Instagram campaigns. To reap full benefits of this comprehensive knowledge of at least 30 platforms, instructors of classes with fewer students than platforms needed to pick up the slack and cover the remaining tools. Finding ways to talk about the platforms that were not covered by students could be challenging. However, on the other side of the spectrum, larger classes may find it challenging to find enough platforms to be covered by every student, in addition to having the appropriate amount of class time allotted to go through all of the platform presentations. Others using this setup should consider and brainstorm the potential for other ways of covering platforms not initially chosen by students for boot camp.

Foster Continued Reflection. Students had an online classroom space where they were encouraged to share their boot camp presentations with one another. This allowed them to use the presentations as a reference as they worked on assignments throughout the semester. This space also had a discussion thread feature; encouraging continued reflection and organic discussion via such an online discussion board during and after boot camp can help keep cyclical experiential learning in process. Building and rebuilding knowledge, as a process rather than an outcome, is key for experiential learning. Ensuring that content from boot camp is integrated into subsequent class assignments, discussions, and experiences in ways that encouraged both conceptual abstraction and practical application/experimentation would be integral for students to continue to build knowledge.

Know the Platforms. Grading the presentations required that the instructor was aware of some of the basics behind each platform from the beginning to assess whether students presented the information that best fit assignment criteria. Sometimes presentations covered only part of what a platform was capable of, or the student did not find a positive example of how the platform could be used in public relations. At those times, there was discussion between presentations regarding how the platform could be seen in relation to the others discussed and whether/how other students were familiar with its capabilities. Often, the necessary material could be gleaned and discussed with the class before moving on to the next platform.

Create Opportunities for Real-World Application. Encouraging students to apply strategic use of presented platforms to the practice of public relations in general was important for experiential learning, as was requiring students to reflect on use for specific clients/organizations to help to deepen understanding of the options and creativity various platforms make available. Students benefited from opportunities to think about how particular platforms could be useful and to make the decisions between platforms, requiring careful consideration of what they knew about both the platforms and the goals of the client. The challenge with this, however, was encouraging students to "think outside the box" and attempting to engage platforms outside of the widely known Facebook/Twitter/Instagram. Students were hesitant to stray from what they already knew, and sometimes those who decided to take a risk and try something different sometimes had less-than-desirable results, which both disappointed and discouraged students from trying something new.

Discuss Downsides. As a large part of the boot camp assignment entailed students discussing uses of platforms for public relations purposes, a risk would be that students leave the exercise with an artificially rosy sense of social media's possibilities and successes. Instructors should ensure that students consider the negative aspects of social media as well as the benefits. For just one example, although students noted that using social media for an organization could give them information on demographics and psychographics of key publics, they did not yet demonstrate understanding the extent and consequences of the digital shadows they and their organizations leave behind. In teaching social media, and in teaching students how to mine the Internet for information, it is important to discuss what information is stored, where it exists, and how it can be found and used in relation to various platforms—as well as the related ethical implications.

DISCUSSION

Data analysis revealed that the new media boot camp teaching approach helped students advance understanding about the practice of public relations at it relates to social media while fostering and embracing conditions for connectivism and experiential learning. Furthermore, five best practices for implementing similar assignments/approaches were identified. Overall, this study produced a number of theoretical and practical implications.

Theoretically, through the fast-paced, information-packed sessions, students were immersed and engaged in grasping information about social media for public relations; and through reflective memo writing, they transformed their boot camp experiences. Throughout the semester, they were able to revisit the knowledge base created at the beginning of the semester through boot camp activities, adding new perspectives, challenging previous assumptions, and re-creating knowledge together. This process of concrete experience, abstract conceptualization, reflective observation, and active experimentation (Kolb, 1984) enabled students to engage in experiential learning, with emphasis on connectivism.

Connectivism in particular was helpful in understanding boot camp as it allowed the students to acknowledge where they were using their networks to gain and gather information that they might not otherwise keep. Students who were willing to discuss which platforms they were familiar with, and which they knew less about, were often talking about platforms where it was easier to acknowledge that the platform existed, and then to let someone (or something, such as a search engine or another online resource) keep that information for them. They knew where to go to gather the information, as they were able to craft and present successful presentations, and they used the classroom network to share that information with their future colleagues, two key skills discussed in connectivism. Additionally, the boot camp assignment allowed students to harness the full potential of the Internet as they learned how technology builds community, both personally and professionally, which provided a notable example of how a young public relations professional can discern and work between the two.

From an applied stance, as noted above, public relations practitioners are highly sought for their new media knowledge and expertise, but there may be a disconnect between today's students'—the so-called digital natives—perceptions of personal social media knowledge and actual professional proficiencies (Melton & Hicks, 2011; Todd, 2009). Boot camp helped combat this disparity by urging students to cycle through the experiential learning process to realize on their own that their self-perceptions about personal new media mastery might not jibe with the professional and rapidly changing new media landscape. Through both concrete and reflective experiences, students began growing their knowledge base not only for nuts-and-bolts skills acquisition but also for integrative public relations activities. Students showed increased critical thinking, especially related to devising social media strategy for particular clients/organization/problems, recognizing benefits and drawbacks of specific tools/strategies, and finding and understanding publics.

In memos, students provided many suggestions for how a particular client or organization could use some of these platforms strategically. Engaging in public-relations-oriented creative and critical thinking early in the semester then carried through the rest of the course. Through additional assignments such as creating social media campaigns for particular clients or conducting an organizational social media audit, students more thoroughly strategized about and logically supported the choices and recommendations they made. However, other than the constantly shifting nature of social media and the popularity of platforms, the students were less open about the challenges they could potentially face as practitioners. The inundation of social media in the students' daily lives could be influencing how they view the platforms overall; they were often more willing to ignore the ability of these platforms to cause or spur issues (and even crises) for organizations and only emphasize the benefits.

Finally, five best practices for implementing similar assignments were offered here: maximizing the number of platforms covered, facilitating sustained reflection, staying abreast of the latest social media platform offerings, offering real-world applications, and remembering to cover downsides as well as benefits of social media usage. In sum, boot camp allowed students to take a more holistic view of how a variety of social media platforms enhance the work of public relations practitioners. By including aspects of experience and networking as integral parts of the learning process, boot camp brought students together to share ideas and experiences in ways that offered positive impacts on their learning. Students and instructors alike can look to the boot camp assignment as a worthwhile way to better understand and maintain knowledge of the benefits and drawbacks of an ever-changing cast of social media tools and applications.

REFERENCES

Alexander, D. (2004). Changing the public relations curriculum: A new challenge for educators. *PRism, 2.* Retrieved from http://www.prismjournal.org/fileadmin/Praxis/Files/Journal_Files /Issue2/ Alexander.pdf

Baker, A. C., Jensen, P. J., & Kolb, D. A. (2002). *Conversational learning: An experiential approach to knowledge creation.* Westport, CT: Quorum.

Benecke, D. R., & Bezuidenhout, R.-M. (2011). Experiential learning in public relations education in South Africa. *Journal of Communication Management, 15*(1), 55–69.

Bonzo, J. (2010). The paradox of social media and higher education institutions. *Proceedings of the 7th international conference on networked learning.* Retrieved from http://www.lancs.ac.uk/fss/ organisations/netlc/past/netlc2010/abstracts/PDFs/Bonzo.pdf

Briones, R. L., & Toth, E. L. (2013). The state of PR graduate curriculum as we know it: A longitudinal analysis. *Journalism & Mass Communication Educator, 68*(2), 119–133.

Brown, B., Sikes, J., & Willmott, P. (2013). *Bullish on digital: McKinsey global survey results.* McKinsey & Company. Retrieved from http://www.slideshare.net/GaldeMerkline/mckinsey-bullish-on-digital

Brown, J. S. (2002). Growing up digital: How the web changes work, education, and the ways people learn. *United States Distance Learning Association Journal, 16*(2). Retrieved from http:// www.usdla.org/html/journal/FEB02_Issue/article01.html

Commission on Public Relations Education. (2006). *The Professional Bond.* Retrieved from http:// www.commpred.org/theprofessionalbond/

Corbin, J., & Strauss, A. (2008). *Basics of qualitative research* (3rd ed.). Thousand Oaks, CA: Sage.

Downes, S. (2007). *What connectivism is.* Half an hour blog. Retrieved from http://halfanhour.blogspot.co.uk/2007/02/what-connectivism-is.html

Ebben, M., & Murphy, J. S. (2014). Unpacking MOOC scholarly discourse: A review of nascent MOOC scholarship. *Learning, Media & Technology, 39*(3), 328–345.

Fitzsimmons, T. (2008). *Exploring connectivism and its effect on training transfer.* Paper presented at the National Communication Association annual conference, San Diego, CA.

Fraustino, J. D., Briones, R., & Janoske, M. (2015). "Can every class be a Twitter chat?": Cross-institutional collaboration and experiential learning in the social media classroom. *Journal of Public Relations Education, 1*(1), 1–18.

Gonzalez, C. (2004). *The role of blended learning in the world of technology.* Retrieved from http://www.unt.edu/benchmarks/archives/2004/september04/eis.htm

Kolb, D. A. (1984). *Experiential learning: Experience as the source of learning and development.* Englewood Cliffs, NJ: Prentice-Hall.

Kolb, A. Y., & Kolb, D. A. (2005). Learning styles and learning spaces: Enhancing experiential learning in higher education. *Academy of Management Learning & Education,* 4(2) 193–212.

Kolb, A. Y., & Kolb, D. A. (2009). Experiential learning theory: A dynamic, holistic approach to management learning, education and development. In S. J. Armstrong & C. Fukami (Eds.), *Handbook of Management Learning, Education and Development.* London: Sage.

Locker, K. O., & Kienzler, D. (2012). *Business and administrative communication* (10th ed.). Boston, MA: McGraw-Hill/Irwin.

Lombardi, A. (2013, December 16). *Social media jobs get harder to fill.* Retrieved from http://www.wantedanalytics.com/insight/tag/skills-in-demand/page/2/

Melton, J., & Hicks, N. (2011). Integrating social and traditional media in the client project. *Business Communication Quarterly, 74*(4), 494–504.

Moody, M., & Bates, E. (2013). PR students' perceptions and readiness for using search engine optimization. *Journalism & Mass Communication Educator, 68*(2), 166–180.

Siemens, G. (2005). Connectivism: A learning theory for the digital age. *International Journal of Instructional Technology and Distance Learning, 2*(1).

Solis, B., & Li, C. (2013). *The state of social business 2013: The maturing of social media into social business.* Altimeter. Retrieved from http://www.slideshare.net/Altimeter/report-the-state-of-social-business-2013-the-maturing-of-social-media-into-social-business

Stone, Z. (2013, July 29). *Social media school is on the rise—but is it necessary?* Retrieved from http://blogs.wsj.com/digits/2013/07/29/social-media-schooling-is-on-the-rise-but-is-it-necessary/

Thompson, T., Jr., Hertzberg, J., & Sullivan, M. (2013). *Social media risks and rewards.* GrantThorton. Retrieved from http://www.grantthornton.com/~/media/content-page-files/advisory/pdfs/2013/ADV-social-media-survey.ashx

Todd, V. (2009). PRSSA faculty and professional advisors' perceptions of public relations curriculum, assessment of students' learning, and faculty performance. *Journalism & Mass Communication Educator, 64,* 71–90.

Waters, R. D., & Bortree, D. S. (2011, Spring). Communicating with millennials: Exploring the impact of new media on out-of-class communication in public relations education. *Teaching Public Relations Monograph, 80.*

Wilson, B. (2012). Service-learning in the public relations classroom: An experiential approach to improving students' critical-thinking and problem-solving skills. *Teaching Public Relations Monograph, 83.*

Wright, D. K., & Hinson, M. D. (2010). How new communications media are being used in public relations: A longitudinal analysis. *Public Relations Journal, 4*(3), 1–27.

ASSIGNMENT 6.1. NEW MEDIA BOOT CAMP

Boot camp is designed to help you learn about a wide variety of social media technologies, how they work, and how or why they might best be used by communication professionals. The assignment has two main requirements: a presentation to the class about a specific platform, and a reflective memo after boot camp is completed.

Presentation

You will research a media website, platform, tool, or app from the list below and present your findings. Your presentation should be **five to seven minutes,** with brief time for Q&A after the presentation. Presentations will be timed. All students will present in class. Topics and presentation days will be determined in class.

Each presentation must include the following topics:

1. What is the new media technology?
   a. Background of the technology—who started it, where did the idea come from, how does it fit into the current social media landscape?
   b. Is there any cost associated with it?
2. Personal/professional use
   a. How you and your classmates can use the new media technology for personal/professional use—and the differences between the two
3. How to sign up/create an account, including important potential settings and/or privacy settings that can be changed or adjusted
4. Best practices and strengths and weaknesses of the platform from a business or communication practitioner standpoint

Be creative and interactive! You must incorporate multimedia, new media, and/or presentation technology in some way (PowerPoint, Prezi, SlideShare, YouTube, podcasts, handouts, activities).

The presentation should cite other sources/evidence. Evidence gives you credibility, and citing it gives credit where it is due. You may include course concepts discussed in class, course readings, and outside sources (academic, trade, or popular press sources). Use APA style for citing.

In terms of content, this is just an overview. You cannot cover everything in five to seven minutes, so choose what is most important/informative (and what best meets the criteria above). Try to give the class a top-level, practical understanding of the technology.

Reflective Memo

Turn in your memo via hard copy at the start of the class period two weeks after your presentation. It should be at least two pages in length, double spaced, Times New Roman 12-point font.

The memo should include a discussion of which platform(s) you thought were most helpful or interesting or unique from a communication practitioner's perspective and why. It should also discuss ways in which you think a particular organization of your choice could use the platforms you like best, along with how and why those platforms might work for that organization as a part of a thoughtful media strategy.

The memo should cite credible sources/evidence. Evidence gives you credibility, and citing gives credit where it is due. You may include course concepts discussed in class, course readings, and outside sources (academic, trade, or popular press sources). Use APA style for citations.

Social Media Technologies

You may choose from the following list of new media technologies for your social media platform (the specific technology you will present on is in italics). You will sign up in class. You are strongly encouraged to choose a platform you are not familiar with to broaden your understanding.

1. Social Network: *Facebook;* 2. Social Network: *Google+;* 3. Social Network: *Meetup;* 4. Professional Social Network: *LinkedIn;* 5. Professional Social Network: *SlideShare;* 6. Professional Social Network: *Scribd;* 7. Blogging: *Squarespace;* 8. Blogging: *Typepad;* 9. Blogging: *Blogger;* 10. Lifecasting: *LiveStream;* 11. Lifecasting: *Vivo;* 12. Podcasting: *iTunes podcasts;* 13. Microblogging: *Twitter;* 14. Microblogging: *Tumblr;* 15. Microblogging: *Storify;* 16. Location-based app: *Foursquare;* 17. Social Search: *SocialMention;* 18. Social Search: *Technorati;* 19. Photo Sharing: *Pinterest;* 20. Photo Sharing: *Picasa;* 21. Photo Sharing: *SnapChat;* 22. Photo Sharing: *Imgur;* 23. Photo Sharing: *Instagram;* 24. Photo Sharing: *Flickr;* 25. Social Bookmarking & Sharing: *Digg;* 26. Social Bookmarking & Sharing: *Reddit;* 27. Social Travel Networks: *TripAdvisor;* 28. Social Commerce: *Groupon;* 29. Social Commerce: *Living Social;* 30. Product/Company Reviews: *Yelp;* 31. Product/Company Reviews: *Angie's List;* 32. Product/Company Reviews: *BizRate;* 33. Management & Measurement: *Hootsuite;* 34. Management & Measurement: *Klout;* 35. Management & Measurement: *TweetDeck;* 36. Video Sharing: *YouTube;* 37. Video Sharing: *Vimeo;* 38. Video Sharing: *Daily Motion;* 39. Video Sharing: *Vine;* 40. Social Q & A: *Wiki Answers;* 41. Wikis: *Wikipedia;* 42. Wikis: *Scholarpedia;* 43. URL Shortener: *tinyurl;* 44. URL Shortener: *bitly;* 45. E-Commerce: *Shopify.*

CHAPTER SEVEN

The Art of Tweeting

Integrating Primary Social Media Research into a Public Relations Writing Course

BETH L. SUNDSTROM AND ABBEY BLAKE LEVENSHUS

> A mediated public relations course must start from the well-documented assumption that the student—indeed, most citizens—have a limited understanding of technology. That is, just because you can surf the Web does not mean that you understand how it is 'surfing' you. (Kent, 2001, p. 60)

Today's undergraduate public relations writing class is undergoing an identity crisis. Educators are tasked with determining what skills to teach amid conflicting opinions. Industry and trade publications included divergent headlines about the practice of public relations such as this one: "'The press release is dead,' Declares the Government's Comms Chief Alex Aiken" (Magee, 2013, p. 3) and this one a month later: "Reports of the Death of the Press Release Have Been Greatly Exaggerated" (Magee, 2013, p. 1).

The Accrediting Council on Education in Journalism and Mass Communications (ACEJMC) updated its curriculum and instruction standards in September 2013 to help prepare students to work in a diverse global and domestic society (ACEJMC, 2013). One of the council's competencies includes applying current tools and technologies and understanding the digital world as it relates to public relations. The accrediting council seeks syllabi and course material evidence that students are given instruction that enables them to gain "digital, technological and multimedia competencies in keeping with professional expectations" (ACEJMC, 2013). To meet these standards, some programs integrate social and digital media across the curriculum. Other schools have added specific courses related to digital tools, technologies, and strategies. Some schools have taken a hybrid approach, fully integrating digital and social tools and content across curriculum and also offering relevant required or elective courses that delve deeper into one aspect, such as writing for digital and social media. This integrated advanced writing class approach may help meet another ACEJMC standard—teaching students to write in forms and styles appropriate to public relations and its audiences and purposes.

Public relations educators have a responsibility to stay current on the research indicating *how* engaging with publics online may differ from offline interactions, *what* best practices are for online communication, and to teach students *how* to strategically craft and tactically execute an effective online writing program for organizations. The ACEJMC standards also call on educators to teach students how to conduct research and evaluate information using appropriate methods to public relations and to help students understand concepts and apply theories related to using and presenting images and information. Traditional public relations writing tactics remain important; one strategy to meet these standards is to interweave multiple professional values and competencies into such courses.

Along with shifting opinions, standards for curriculum and instruction have evolved to keep pace with the current multimedia landscape that characterizes today's organization-public interactions. It can be difficult to decide what to incorporate and what to omit in order to prepare students well. Thus, this chapter offers one approach addressing several competencies through a collaborative social media research project in an undergraduate public relations writing course that helps teach students to *write* for social media. Instructors must also negotiate the relevance of AP style on Twitter; grammar, spelling, and punctuation basics on Facebook; the pitch letter to bloggers; and the importance of media kits in order to prepare students for the practice of public relations. Specifically, this chapter offers a case study of conducting collaborative primary research as an exemplar approach to teaching social media writing in an upper-level undergraduate public relations class in order to engage students in their own learning of how to craft dialogic messages on Twitter.

LITERATURE REVIEW

Active Learning in Social Media Research

Freire (2005) described the tension between teaching-learning in leading students on a path toward knowing. In public relations pedagogy, active learning would integrate conceptual knowledge with skills training (Aldoory & Toth, 2000; Kolb & Kolb, 2005; Vasquez & Botan, 1999). This approach would facilitate opportunities for learning by doing and would emphasize strategic thinking and problem-solving skills. Active learning would construct a shared community of knowledge in which the instructor would design experiences to foster learning while students would maintain increased responsibility for learning (Berger, 2002). Through these experiences, public relations students would apply theory and concepts in the context of building relationships with publics.

Educators in the applied discipline of public relations have been supporting active learning opportunities, which allowed students to engage with theory in a practical setting (Fall, 2006; Hoag, Jayakar, & Erickson, 2003). Active learning has been playing a fundamental role in developing critical and reflective thinking competencies among undergraduate public relations students (Berger, 2002; Lubbers & Gorcyca, 1997). Strong writing skills, research experience, and the ability to work in teams have been among the most desirable attributes for public relations professionals (Stacks, Botan, & Turk, 1999).

Yet, digital media have broadened and complicated public relations writers' work (McEachern, 2013) and posed pedagogical challenges and opportunities for media and communication instructors (Moody, 2010; Platt, 2011). While public relations and other mass communication students still need solid traditional media writing skills such as press releases, they are also now asked to engage with online stakeholders and build online communication campaigns and strategies, including on platforms such as Twitter (Platt, 2011). As such, instructors today must integrate new technologies into their curriculum so that students can "develop the digital literacy skills they will need to succeed in the twenty-first century workplace" (Platt, 2011, p. 229).

In addition to teaching writing, public relations educators have emphasized the importance of teaching students how to conduct public relations research (King, 2001) and how to use new technology to do so (Kent, 2001). Frequently, undergraduate writing courses include case studies or planning elements, but few include primary research (King, 2001). Engaging in applied social media primary research emphasizes creativity and allows students to bridge the gap between theory and application. Existing research has also identified a gap between professionals' recognition of social media as expanding organizations' ability to engage with online audiences and their actual limited use of social media (Esrock & Leichty, 1999; Kent, Taylor, & White, 2003; Seltzer & Mitrook, 2007). This project, therefore, used active learning to help students identify research-based social media writing best practices and then critically analyzed actual use in the context of media companies' Twitter activity. Twitter was particularly useful as an active, applied learning platform for students due to its interactive characteristics, such as the ability for senders and receivers to post messages and post-update and categorization tools such as hashtags, keywords, and the option to follow and retweet, or repost, other users' tweets (Smith, 2010).

Media Companies and Twitter Engagement

While scholars have identified Twitter-specific aspects for interactivity, many organizations fail to use the online micro-blogging platform to its maximum engagement potential and instead use it mostly to distribute information and messages (Linvill, McGee, & Hicks, 2012; McAllister, 2012; Waters & Jamal, 2011). One of the benefits of more effectively using social media include improving organization-public relationships (Saffer, Sommerfeldt, & Taylor, 2013). Saffer et al. (2013) found that corporations with more highly interactive Twitter feeds improved audience perceptions of organization-public relationships compared to less active organizations. While research has identified positive benefits of effective online communication with publics, few studies have focused on strategies that organizations could use to interact with publics on social media platforms such as Twitter (Saffer et al., 2013).

Communication strategies, whether planned or emergent, can guide organizational communication with audiences and stakeholders (Steyn, 2007). In addition to preplanned strategies, some emerged as organizations adjusted to ongoing issues, events, opportunities, and crises (Steyn, 2007). In a study of *Fortune* 500 corporate blogs, Lee, Hwang, and Lee (2006) found that companies leveraged product development and customer service, thought leadership, and promotion content strategies. While most companies used top-down controlled approaches where thought leadership and promotional strategies aimed to shine a positive light on the company, Lee et al. (2006) found some companies used a bottom-up thought leadership approach that included disseminating individuals' opinions and insights about the company (Lee et al., 2006).

A similar bottom-up approach or strategy within a Twitter context included retweeting others' content to foster a dialogic or conversational loop (Linvill et al., 2012). To boost engagement and relationship-building, organizations could also use individual and group linking by naming individuals or organizations within a tweet or retweet using Twitter handles such as @username (Smitko, 2012). Linking to credible, external sources could help provide useful information and generate return visits—two key dialogic principles (Linvill et al., 2012). Organization type, size, and prominence could influence amount of tweetable external content such as media coverage or awards.

As such, top media companies could hold Twitter content advantages based on size, prominence, and variety of content they produce. At the same time, media companies face unique expectations from stakeholders about online engagement (Mersey, Malthouse, & Calder, 2010). Social media platforms, including Twitter, offer consumers increased agency in their interactions with organizations. Media companies must adapt new technologies to effectively disseminate content to and engage with target audiences (Mersey et al., 2010). Media companies include film, music, broadcast radio, television, publishing, live entertainment, advertising, and sports, among others. Historically, many of these organizations served as content producers, distributors, and broadcasters to passive consumers. Today, they have been operating in an online environment of bidirectional, multimedia communication, with low entry barriers (Hess, 2014). Media companies operating in this context may hold content advantages, yet shifting stakeholder expectations influence their online engagement (Mersey et al., 2010). To build on existing research within the unexplored media company context, we asked: How do media companies' tweets incorporate dialogic strategies that foster engagement?

METHOD

New Media Writing and Strategic Communication: An Advanced Writing Course. This was an advanced writing course emphasizing public relations practices. In this course, students would improve their writing skills and would gain an understanding of theoretical and best practice resources for communicating well. The course introduced students to the use and influence of new media in public relations practice and expanded their ability to write using social media platforms and tools. Students would gain critical analysis skills to effectively critique organizations' media writing and use. This course offered students the opportunity to produce quality written materials for a portfolio of writing and analysis.

Throughout the semester, every member of the course was involved in an iterative research project. The assignment offered students an opportunity to be involved in the conduct of a new media research project. The theoretical goal of the assignment was to create new knowledge in the field of public relations. The practical goal of the assignment was to better understand how to write effective messages online. Meanwhile, the active learning assignments included out-of-classroom projects, team assignments, reflection, applied research, and technologies (Berger, 2002).

The research project was worth 20% of the overall course grade. The following characteristics were assessed in grading students' written work throughout the semester: accuracy of content (evidence of strategic, in-depth thinking and analysis, especially related to social media principles); mechanics (e.g., grammar, spelling, punctuation); organization of ideas and writing quality; conformance to AP format, style guides, standards covered in class, and assignment instructions; creativity; and professional appearance.

Bringing a Social Media Research Project into the PR Writing Classroom. In this course, students were introduced to social media writing best practices. While research highlights the importance of two-way communication online, little is known about how to create dialogue with stakeholders. To address that gap, students helped develop a study exploring how the dialogic theory of public relations (Kent & Taylor, 2002) could help practitioners improve dialogic engagement between organizations and publics.

Students investigated Twitter communication from 18 top media companies, examining aspects of dialogic communication and suggesting evidence-based best practices for social media writing. They engaged on Twitter and Google Drive to conduct a qualitative content analysis, including literature review, data collection, coding, data analysis, identifying themes and patterns across data, drawing conclusions, and discussing theoretical and practical implications.

APPLICATIONS

The social media research project involved a qualitative content analysis to investigate dialogic communication online. The dialogic theory of public relations (Kent & Taylor, 2002) provided a conceptual framework to explore how strategic communication practitioners foster dialogic engagement online between organizations and publics (Linvill et al., 2012; Rybalko & Seltzer, 2010). Student responsibilities were incorporated at the individual, team, and class levels. Students were accountable for individual and team-based tasks to enrich learning outcomes. Critical reflections, including individual discussion board posts, group sessions, and research writing assignments, provided opportunities for students to interrogate new information and concepts. These reflections were shared with peers online and during class discussions, stimulating new understandings of course concepts.

Why Twitter? Twitter plays an increasingly important function in the public relations of organizations and governments. Public relations scholars identified and explored the Internet's potential to improve dialogic communication between organizations and publics (Hallahan, 2008; Kelleher, 2009; Kent & Taylor, 1998, 2002). Twitter allows public relations practitioners to disseminate content and engage with target audiences, while simultaneously raising the voices and power of consumers. The online microblogging service Twitter was used to identify tweets released by organizations for data analysis.

In conducting this research project, students were required to negotiate organizational Twitter handles, profiles, and accounts. There was a significant value in such experience because students must be proficient in social media platforms to succeed as public relations professionals. This social media research project integrated Twitter into pedagogy because students must understand social media platforms they would be required to use in professional practice.

Many students were not familiar with Twitter at the start of the course, and even those with a personal account were unfamiliar with professional and organizational applications. To address this knowledge gap, the instructor started a discussion board including "how-to" resources where students could ask each other questions about the platform and engage in problem solving about required tasks. Students were also able to ask one another questions on Google Drive by making a comment in a document or spreadsheet. The instructor received an email notification of all comments and responded if other students did not answer the question appropriately in a timely manner. In most cases, students resolved problems and confusion within 24 hours. Since Twitter was required for this course, a new media and technology policy was included in the syllabus.

Guiding Students Through Extant Theoretical and Conceptual Literature

This course incorporated readings on the role of new media in the following public relations functions: relationship management; research and environmental scanning; publics; media/blogger relations; organizational voice; internal communication; measurement and evaluation (metrics); issues and crisis management; law and ethics; and diversity. These theoretical and conceptual areas situated the course and project within the larger public relations research sphere. The instructor located the social media research project within a relationship management and organization-public relationship framework. Therefore, as a component of required readings, a literature review was conducted to learn more about relationship management, communication approaches that were used by practitioners, and previous research on the topic that were discussed in class to establish the state of current thinking related to media companies' engagement on Twitter. Although training in research methods was a prerequisite for this course, students benefitted from a step-by-step process involving in-class trainings and out-of-class homework assignments. A minimum of five course periods were reserved for research project workshops with additional discussions throughout the semester.

The research project spanned the length of the semester and provided opportunities for students to identify real-world and academic connections to the required readings (see Assignment 7.1). The research project was broken down into the following steps.

1. Google Drive. Prior to the start of the research project, students were trained to use Google Drive. Cmail is the college's student email system powered by Google. All students already had a @g.cofc.edu email account. Through the college's partnership with Google, students also maintained access to Google Apps, Google's set of hosted and customizable communications solutions, including Google Drive. All services were hosted by Google and were available to users from any smartphone, tablet, or computer. The class used the Docs and Sheets apps as the research project database.

2. Sampling Workshop. Through a rigorous sampling process, 18 for-profit media and entertainment companies were chosen for the study. These organizations were chosen from SNL Financial's list of the top 25 media and entertainment companies based on net income from the second quarter of 2012 (Sadiq & James, 2012). SNL Financial provided financial data and expert analysis on business sectors critical to the global economy, including media/communications. Only those organizations with a Twitter handle and a regularly updated Twitter feed were studied. Following Linvill et al.'s (2012) definition of "active," researchers visited Twitter to determine which of the top 25 media companies had tweeted within one week of the designated date (January 25, 2013). Of the 25 top media companies, 18 were found to be active and were included in the study. See Table 7.1 for a list of the organizations studied.

Table 7.1. Media Organizations Sample from SNL Financial's Ranking Based on Q2'12 Net Income

Organization	Twitter Handle	Twitter Bio
Walt Disney Co.	@Disney	"We seek to estimate the future and its bearing on our existence, as well as dwelling fondly on the past or indulging in escapist dreams."—Walt Disney facebook.com/Disney
Thomson Reuters Corp.	@thomsonreuters	"The world's leading source of intelligent information for businesses and professionals. This account is managed by our Social Media team." Worldwide. thomsonreuters.com
Viacom Inc.	@Viacom	"Proud parent of BET, Centric, CMT, Comedy Central, Logo, MTV, Nickelodeon, Paramount Pictures, Spike, TV Land, VH1 and Viacom International Media Networks." 160+ countries and territories. viacom.com
Time Warner Inc.	@twxcorp	"The official Twitter account for Time Warner Inc. (NYSE:TWX)" New York, NY. timewarner.com
CBS Corp.	@CBSTweet	"Official twitter for CBS Television Network." cbspressexpress.com

Discovery Communications Inc.	@DiscoveryComm	"News from Discovery Communications (NASDAQ: DISCA, DISCB, DISCK), the world's number one nonfiction media company" Silver Spring, MD blog.discoverycommunications.com
McGraw-Hill Cos. Inc.	@McGrawHillCos	"A global financial and education company."
Scripps Networks Interactive Inc.	@scrippscomm	"As the leader in lifestyle media, our brands—HGTV, DIY Network, Food Network, Cooking Channel, Travel Channel & GAC—connect w/viewers every day, everywhere!" Knoxville, TN. blog.scrippsnetworks.com
Gannett Co. Inc.	@Gannett	"We're tweeting about trends in media marketing and what's happening with our companies. Mobile, Publishing, Broadcast, Digital." gannett.com
Interpublic Group of Cos. Inc.	@InterpublicIPG	"Interpublic is one of the world's premier advertising and marketing services companies with 44,800 employees in over 100 countries. (NYSE: IPG)" New York, NY. interpublic.com
Nielsen Holdings NV	@Nielsen	"Global consumer and media insights from Nielsen." New York, NY. nielsen.com
Six Flags Entertainment Corp.	@SixFlags	"Follow your favorite local Six Flags park Twitter page for news, exclusive discounts, and contests!" 18 locations in North America. sixflags.com
Super Media Inc.	@SuperMedia	"We help make local businesses successful in local marketing. #seo #sem #socialmedia #print #mobile Visit our Small Business Advisor blog: http://spag.es/L1qP2t" Dallas, TX. supermedia.com
Scholastic Corp.	@Scholastic	"The official Twitter account of Scholastic, including Scholastic Reading Club and Book Fairs. Tweet us with your thoughts on books & education!" New York, NY. scholastic.com

Dex One Corp.	@DexOne	"Marketing solutions for local business! http://about.me/dexone #smallbiz #seo #sem #socialmedia." USA. dexone.com
Washington Post Co.	@washingtonpost	"Tweets about everything from breaking news to bad restaurants. Sometimes global, sometimes local. Most tweets by @ TJOrtenzi." Washington, D.C. washingtonpost.com
AMC Networks Inc.	@AMCnews	"On-air and online reports about AMC's original series and entertainment news." Basic Cable. amcnews.com
Regal Entertainment Group	@RegalMovies	"Regal Entertainment Group operates the largest chain of movie theatres in the US. Follow us for giveaways, movie news, and so much more!" Across the United States. regmovies.com

Updated November 1, 2013

Note. Sadiq, D., & James, S. B. (2012, Sept. 19). News Corp. falls from list of top media earners, based on bottom line. Top 25 media and entertainment companies by Q2'12 net income. Data Dispatch. SNL Financial, Charlottesville, VA. Retrieved from https://www.snl.com/InteractiveX/Article.aspx?cdid=A-15838947-10792

3. Data Collection. To ensure that an individual company did not dominate the sample, 100 tweets from each organization were identified. Each student was assigned one organization. Starting on January 25, 2013, the preceding 100 tweets were collected and entered into a Sheet on Google Drive. In addition to collecting each tweet's content, students recorded the number of retweets, favorites, and replies, so that each organizational tweet was analyzed with context about stakeholders' interactivity and engagement with the message. A total of 1,800 tweets from 18 media organizations were collected and entered into Google Drive. Data collection was an out-of-class homework assignment.

4. Coding Workshops. In this study, dialogic theory provided a conceptual framework to develop initial guidelines for coding the tweets. Following other dialogic theory studies (Linvill et al., 2012; Rybalko & Seltzer, 2010), students learned about and investigated the following dialogic principles: useful information, conservation of visitors, generation of return visits, and the dialogic loop. The study employed analysis techniques drawn from three concurrent activities: data reduction, data display, and conclusion drawing and verification (Miles & Huberman, 1994). In class, students proposed a codebook based on concepts from dialogic theory and the research question to maintain consistency in coding. Students were trained to employ an iterative process of data reduction, using descriptive, interpretive, and pattern codes to identify themes across the data. In class, students learned how to use a codebook and practiced coding.

Following this process, students coded tweets independently and met to resolve any differences in coding and to adapt the codebook. Google Drive Excel provided a living document where researchers could view each other's coding and make comments in real-time. Google Drive also served as an interactive data display to present information systematically.

5. Check-Coding. According to Miles and Huberman (1994), check-coding helped attain definitional clarity and served as an appropriate check of coder consistency. When at least two-thirds of the coding was completed, students learned about the principle of intercoder reliability or interrater reliability. Students learned various methods to examine the agreement between coders in a content analysis when multiple trained coders categorize text. The discussion emphasized that this is one way to measure the implementation and accuracy of a codebook or coding system. In class, students employed the Miles and Huberman (1994) check-coding formula.

6. Qualitative Data Analysis Workshop. Students learned that qualitative data consists of words and stories requiring creativity and a systematic approach to organize and make meaning from the data. As with all research, there are many approaches to analysis and interpretation. To prepare for an in-class research team meeting, students were instructed to:

I. Immerse themselves in the codes.
   a. Read and reread the quotes in the context of the codes.
   b. Write down initial impressions.

II. Identify themes or patterns in the data.
   a. Look at the tweets with the most retweets, favorites, replies. Ask if they have anything in common.
   b. Sort columns by each code and read all of the quotes. Do these quotes have anything in common? (For example, what types of words are used as hashtags—nouns, verbs, names, and so on?)
   c. Write down themes or patterns (such as concepts, terms, phrases, interactions, people, relationships, questions, and so forth).

III. Write a discussion board post reflecting on your initial analysis of the data.
   a. Be sure to consider the research question.
   b. Provide thoughtful feedback and comments on your peers' analysis and identify areas of similarities and differences with your own post.

7. Themes/Results Research Discussion. Students learned how to write the results around a theme and each student was assigned one theme to summarize for the study's findings. The out-of-class assignment tasked students with developing an in-depth analysis, rich with description, context, and direct quotations so that each organization was represented in its own terms. Students submitted each theme on one Google Doc to create a singular Results section. Students read and commented on all themes, providing an in-depth constructive peer review for one theme.

8. Conclusions and Implications Discussion. Students learned conclusion drawing and verification techniques to identify themes that answered the research question. They identified unique and unexpected patterns and sought outliers and rival explanations for conclusions. Students also discussed the research in its entirety, including limitations and strengths of the methodology. Specifically, the following concepts were discussed: implications (for theory and practice), limitations (of the study/research methods), and future directions (for additional research). To develop conclusions, students followed the Miles and Huberman (1994) recommendations to revisit findings as a whole, determine how findings answer the research question, and uncover implications of the results. Practical applications from the findings were developed. Students learned to avoid attempts to generalize to other populations or situations by translating results into implications for action.

Following the completion of the course, researchers' additional analyses revealed that media companies used strategies based on dialogic constructs of useful information, conservation of visitors, generation of return visits, and a dialogic loop. A brief summary of results based on further analyses is included here: promotion as thought leadership, social media integration, and synergistic organizational voice.

Promotion as Thought Leadership. Organizations promoted useful information, which established them as thought leaders in the industry. Information useful to publics included news releases, online events, and statistics. Online events included Twitter parties and live chats with celebrities, as well as webinars. Online events employed hashtags and introduced third-party endorsements. Statistics grabbed attention and interest on Twitter, generating conversations with followers. Time Warner (2012, November 5) stimulated discussion with the tweet: "Did u know that 54% of ppl under 30 prefer texting someone over talking to them? Findings from @TWXCorp medialab study: http://bit.ly/Pyt."

Social Media Integration. Social media integration increased conservation of visitors and generated return visits. Tweets that inspired engagement, such as retweets and favorites, were concise, specific, and provided hyperlinks to more information. Dialogic tweets connected publics to the organizations' other social media platforms, including blogs, Facebook, Instagram, YouTube, Foursquare, and Pinterest. Incorporating imagery at strategic times prompted audience engagement on a greater scale and facilitated integration across platforms. Vine and YouTube were the most widely used video sharing services. Organizations used videos to promote products and services, as well as to engage in pressing social issues. Thomson Reuters (2013, January 25) used video to engage in a discussion about the cost of higher education, asking, "Are student loans the next subprime crisis?" with a link to a YouTube video.

Synergistic Organizational Voice. Synergistic organizational voice, which included conversations and interactivity, provided an effective dialogic loop. The ideal tweet length emerged around 100 characters, offering dialogic space for individuals to incorporate comments in retweets. Consistently replying to mentions and answering questions promptly inspired engagement. Organizations conducted online conversations with followers, related organizations, media outlets, and celebrities. These connections created an interactive community of followers around a shared common interest. As an extension of third-party credibility, organizations offered exclusive or quotable content that synergized and extended the organizational voices. Dialogic messages incorporated exclusive content, which included interactive topics, such as quizzes, contests, and free prizes. Hashtags also boosted interactivity by helping start and organize conversations. For example, Disney employed recurring hashtags, such as #GuesstheCharacter, #DisneyTrivia, and #DisneyDecode to connect with publics.

Recommendations for Social Media Writing and Collaborative Research. This section offers final recommendations and challenges related to integrating a collaborative social media research project in an undergraduate public relations writing course to teach students to write for social media. As described above, students demonstrated enhanced learning through the development of evidence-based best practices to improve new media writing. The following Top 10 List was updated based on additional data analyses; however, it was representative of the list developed by the students.

The Top 10 Ways Dialogic Messages Inspire Engagement on Twitter:

1. Tweets of only 100 characters provided space for users to retweet their own comments.
2. Dialogic tweets were concise, specific, and provided hyperlinks to more information.
3. News releases and statistics provided useful information to publics.
4. Dialogic messages incorporated interactive topics, such as quizzes, contests, and free prizes.
5. Hashtags boosted interactivity by helping create and maintain conversations.
6. Mentions, retweets, and replies maintained community and interactivity.
7. Online events, including Live Tweeting, provided a venue to introduce third-party endorsements.

8. Exclusive or quotable content extended organizational voice.
9. Photographs, videos, and other strategic imagery created large-scale audience engagement.
10. Social media integration (such as linking to other new media platforms) enhanced community.

Course assignments, including new media translations, provided opportunities for students to execute their Top 10 List. These evidence-based best practices became the guidepost for the instructor and students to provide feedback on new media writing. Students learned to execute and identify these strategies in organizational new media writing. Students also applied these "writing rules" to assigned course readings, often comparing and contrasting recommendations, noting, "that is what we said!" or "I disagree because our findings showed this . . . instead."

Challenges and Outcomes. While this collaborative social media research project instilled the rigor of new media writing and improved students' writing skills, there were challenges to implementing this assignment. Students expressed concerns about preparation and ability to conduct a rigorous research project. At times, students were overwhelmed by new tasks, such as coding and data analysis. Students expressed frustration when other class members did not fulfill their obligations to the research project. From the instructor's perspective, challenges included development of students' research and technological skills, including Twitter and Google Drive. This assignment may stretch students outside of their comfort zones. Instead of reading, memorizing, and applying information and skills, students are tasked with creating new knowledge. This can be a daunting endeavor for many students and requires patience and dedication from instructor and students.

Student feedback indicated that individual, team, and class accountabilities created a sense of ownership of the class and research project. These collaborative learning strategies mediated interactions between peers that facilitated learning (Lattuca, 2006). The evidence-based best practices and Top 10 List allowed students to make connections between knowledge, theory, and practical application. Through this research project, students were engaged and responsible for learning (Berger, 2002). Students demonstrated enhanced learning and ability to apply these strategies and writing skills. At the conclusion of the course, students valued the applied research experience they gained through this project and demonstrated pride in their rigorous data analysis and findings.

DISCUSSION

Public relations builds relationships between an organization and its publics (Grunig & Hunt, 1984). An active learning course emphasizing public relations practices guided students to improve their writing skills and gain an understanding of theoretical and best practice resources for communicating well (Aldoory & Toth, 2000; Kolb & Kolb, 2005; Vasquez & Botan, 1999). This research project assignment introduced students to the use and influence of new media in public relations practice and expanded their ability to write using social media platforms. Students gained critical analysis skills to effectively critique organizations' media writing and use. Furthermore, this assignment offered students the opportunity to produce quality written materials for a portfolio of writing and analysis.

This research project addressed ACEJMC competencies and values, as well as two of the accrediting body's indicators. First, the assignment integrated theory and skills to achieve a broad range of the Council's competencies. Second, the assignment provided current and challenging instruction responsive to emerging professional expectations related to technological competencies. For example, students were introduced to social media writing best practices. To solidify learning, students were asked to employ evidence-based best practices through in-class new media translation exercises. Students maintained increased responsibility for learning (Berger, 2002) by establishing peer review guidelines to facilitate constructive feedback. Students improved their social media writing skills and became better writers by reading and critiquing one another's work through peer review.

This chapter offered practical strategies for instructors to foster engagement and apply dialogic principles to teach online communication. Dialogic principles enhance transparent organization-public communication and improve public relations (Taylor, Kent, & White, 2001). Scholars suggest that technological advances may require adjustments to dialogic theory (Pettigrew & Reber, 2010). For example, Twitter's inherent interactivity, including retweeting and the use of hashtags, provide an ideal platform to investigate two-way communication (Smith, 2010). This study demonstrated the need to further develop the theoretical and practical applications of dialogic theory and the role of technological platforms in applying these principles.

Through the lens of dialogic theory, this study identified three overarching strategies evidenced in top media companies' tweets: promoting organizations as industry and thought leaders, integrating social media, and employing a synergistic organizational voice. These strategies advanced research explicating the use of social media to create a conversational voice and improve organization-public relationships (Kelleher, 2009; Kelleher & Miller, 2006; Smith, 2010). Practitioners have argued that social media allows every company to be a media company (Brito, 2013; Foremski, 2010). As a result, organizations have the opportunity to create and deliver content directly to audiences. When this content matches audience expectations and needs, social media communication may improve engagement and foster improved relationships.

Kent (2001) argued that students should understand how new technologies offer the potential to improve messages and build organization-public relationships.

This research project emphasized the fit between form and function in online messages. This chapter offered recommendations for conducting collaborative social media research to teach students how to write dialogically on behalf of organizations. Students developed a critical eye, rooted in the academic and industry literature, for how organizations engage interactively online. They reached consensus, through systematic coding and check-coding, for how organizations' tweets integrated dialogic features. Students analyzed overall themes and articulated conclusions and future recommendations. Students used active learning to arrive at their own research-based conclusions. They were participants in their own learning, rather than passive recipients of a bulleted PowerPoint slide about effective online writing. This class project offered a model of a collaborative research study that brings social media into the public relations writing classroom. This chapter contributed to a dialogue among public relations educators demonstrating that primary research would provide an exemplar approach to teaching social media writing.

REFERENCES

Accrediting Council on Education in Journalism and Mass Communications. (2013). ACEJMC-Accrediting Standards. *ACEJMC Accrediting Standards.* Retrieved from http://acejmc.ku.edu/PROGRAM/STANDARDS.SHTML#std2

Aldoory, L., & Toth, E. L. (2000). An exploratory look at graduate public relations education. *Public Relations Review 26,* 115–125. doi:10.1016/S0363-8111(00)00034–5

Berger, B. K. (2002). Applying active learning at the graduate level: Merger issues at Newco. *Public Relations Review, 28,* 191–200. doi:10.1016/S0363-8111(02)00126–1

Brito, M. (2013). *Your brand, the next media company: How a social business strategy can enable better content, smarter marketing and deeper customer relationships.* Indianapolis, IN: Pearson Education.

Esrock, S. L., & Leichty, G. B. (1999). Corporate World Wide Web pages: Serving the news media and other publics. *Journalism & Mass Communication Quarterly, 76,* 456–467.

Fall, L. (2006). Value of engagement: Factors influencing how students perceive their community contribution to public relations internships. *Public Relations Review, 32,* 407–415. doi:10.1016/j.pubrev.2006.09.009

Foremski, T. (2010, March 30). When every company is a media company. EC = MC: The transformative equation for business. Retrieved from http://www.everycompanyisamediacompany.com/every-company-is-a-media-/2010/03/welcome-when-every-company-is-a-media-company.html

Freire, P. (2005). *Teachers as cultural workers: Letters to those who dare teach.* Boulder, CO: Westview Press.

Grunig, J. E., & Hunt, T. (1984). *Managing public relations.* New York, NY: CBS College.

Hallahan, K. (2008). Organizational–public relationships in cyberspace. In T. L. Hansen-Horn & B. D. Neff (Eds.), *Public relations: From theory to practice* (pp. 46–73). Boston, MA: Pearson.

Hess, T. (2014). What is a media company? A reconceptualization for the online world. *The International Journal on Media Management, 16,* 3–8. doi:10.1080/14241277.2014.906993

Hoag, A. M., Jayakar, K. P., & Erickson, K. (2003). The role of trust in virtual and interpersonal environments: Implications for team learning & case method pedagogies. *Journalism & Mass Communication Educator 57,* 370–383.

Kelleher, T. (2009). Conversational voice, communicated commitment, and public relations outcomes in interactive online communication. *Journal of Communication, 59*, 172–188. doi:10.1111/j.1460-2466.2008.01410.x

Kelleher, T., & Miller, B. M. (2006). Organizational blogs and the human voice: Relational strategies and relational outcomes. *Journal of Computer-Mediated Communication, 11*, 395–414. doi:10.1111/j.1083-6101.2006.00019.x

Kent, M. L. (2001). Teaching mediated public relations. *Public Relations Review, 27*, 59–71. doi:10.1016/S0363-8111(01)00070-4

Kent, M., & Taylor, M. (1998). Building dialogic relationships through the World Wide Web. *Public Relations Review, 24*, 321–334. doi: 10.1016/S0363-8111(99)80143-X

Kent, M. L., & Taylor, M. (2002). Toward a dialogic theory of public relations. *Public Relations Review, 28*, 21-37.

Kent, M. L., Taylor, M., & White, W. J. (2003). The relationship between web site design and organizational responsiveness to stakeholders. *Public Relations Review, 29*, 63–77. doi:10.1016/S0363-8111(02)00194-7

King, C. M. (2001). The write stuff: Teaching the introductory public relations writing course. *Public Relations Review 27*, 27–46. doi:10.1016/S0363-8111(01)00068-6

Kolb, A. Y., & Kolb, D. A. (2005). Learning styles and learning spaces: Enhancing experiential learning in higher education. *Academy of Management Learning & Education 4*, 193–212. doi:10.5465/AMLE.2005.17268566

Lattuca, L. R. (2006). The constructivist pedagogy we're looking for. *Journalism & Mass Communication Educator, 60*, 354–358.

Lee, S., Hwang, T., & Lee, H-H. (2006). Corporate blogging strategies of the Fortune 500 companies. *Management Decision, 44*, 316–334. doi:10.1108/00251740610656232

Linvill, D. L., McGee, S. E., & Hicks, L. K. (2012). Colleges' and universities' use of Twitter: A content analysis. *Public Relations Review, 38*, 636–638. doi:10.1016/j.pubrev.2012.05.010

Lubbers, C. A., & Gorcyca, D. A. (1997). Using active learning in public relations instructions: Demographic predictors of faculty use. *Public Relations Review 23*, 67–80.

Magee, K. (2013, September 23). "The press release is dead," declares the Government's comms chief Alex Aiken. *PR Week*. Retrieved from http://www.prweek.com/article/1212883/the-press-release-dead-declares-governments-comms-chief-alex-aiken

Magee, K. (2013, October 10). Reports of the death of the press release have been greatly exaggerated. *PR Week*. Retrieved from http://www.prweek.com/article/1215044/reports-death-press-release-greatly-exaggerated

McAllister, S. (2012). Practitioner perceptions of the importance, function, and actual utilization of dialogic Internet tools and institutional resources at American colleges. *Atlantic Journal of Communication, 20*, 221–236. doi:10.1080/15456870.2012.711151

McEachern, R. W. (2013). Social media challenges for professional writers. *Journal of Current Issues in Media & Telecommunications, 5*, 279–287.

Mersey, R. D., Malthouse, E. C., & Calder, B. J. (2010). Engagement with online media. *Journal of Media Business Studies, 7*, 39–56.

Miles, M. B., & Huberman, A. M. (1994). *Qualitative data analysis: An expanded sourcebook* (2nd ed.). Thousand Oaks, CA: Sage.

Moody, M. (2010). Teaching Twitter and beyond: Tips for incorporating social media in traditional courses. *Journal of Magazine & New Media Research, 11*, 1–9.

Pettigrew, J. E., & Reber, B. H. (2010). The new dynamic in corporate media relations: How Fortune 500 companies are using virtual press rooms to engage the press. *Journal of Public Relations Research, 22*, 404–428. doi:10.1080/10627261003801412

Platt, C. A. (2011). Blogging in the communication technology course. *Communication Teacher, 25*, 228–233. doi:10.1080/17404622.2011.601719

Rybalko, S., & Seltzer, T. (2010). Dialogic communication in 140 characters or less: How Fortune 500 companies engage stakeholders using Twitter. *Public Relations Review, 36*, 336–341. doi:10.1016/j.pubrev.2010.08.004

Sadiq, D., & James, S. B. (2012, September 19). News Corp. falls from list of top media earners, based on bottom line. Top 25 media and entertainment companies by Q2'12 net income. Data Dispatch. SNL Financial, Charlottesville, VA. Retrieved from https://www.snl.com/InteractiveX/Article.aspx?cdid=A-15838947–10792

Saffer, A. J., Sommerfeldt, E. J., & Taylor, M. (2013). The effects of organizational Twitter interactivity on organization–public relationships. *Public Relations Review, 39*, 213–215. doi:10.1016/j.pubrev.2013.02.005

Seltzer, T., & Mitrook, M. A. (2007). The dialogic potential of weblogs in relationship building. *Public Relations Review, 33*, 227–229. doi:10.1016/j.pubrev.2007.02.011

Smith, B. G. (2010). Socially distributing public relations: Twitter, Haiti, and interactivity in social media. *Public Relations Review, 36*, 329–335. doi:10.1016/j.pubrev.2010.08.005

Smitko, K. (2012). Donor engagement through Twitter. *Public Relations Review, 38*, 633–635. doi:10.1016/j.pubrev.2012.05.012

Stacks, D. W., Botan, C., & Turk, J. V. (1999). Perceptions of public relations education. *Public Relations Review 25*, 9–28.

Steyn, B. (2007). Contribution of public relations to organizational strategy formulation. In E. L. Toth (Ed.), *The future of excellence in public relations and communication management: Challenges for the next generation* (pp. 137–172). Mahwah, NJ: Lawrence Erlbaum.

Taylor, M., Kent, M. L., & White, W. J. (2001). How activist groups are using the Internet to build relationships. *Public Relations Review, 27*, 263–284. doi:10.1016/S0363-8111(01)00086-8

Vasquez, G., & Botan, C. (1999). Models for theory-based M.A. and Ph.D. programs. *Public Relations Review 25*, 113–122. doi:10.1016/S0363-8111(99)80132-5

Waters, R. D., & Jamal, J. Y. (2011). Tweet, tweet, tweet: A content analysis of nonprofit organizations' Twitter updates. *Public Relations Review, 37*, 321–324. doi:10.1016/j.pubrev.2011.03.002

ASSIGNMENT 7.1. TWITTER RESEARCH PROJECT

This assignment is worth 20% of the final course grade (200 points). To complete this assignment, you must attend all in-class workshops because these skills build on each other. In addition to course participation, you will be graded on two specific tasks: Coding and Writing the Results/Findings. These two components of the assignment will each be worth half of your overall grade on the assignment (100 points each). This document provides detailed instructions for each component of this assignment. It is your responsibility to check due dates located in the syllabus.

Throughout the semester, every member of this class will be involved in an iterative research project. This project will offer students an opportunity to be involved in the conduct of a new media research project. The theoretical goal of this project is to create new knowledge in the field of communication. The practical goal of this project is to better understand how to write effective messages online. You will be graded on:

- Accuracy of content
- Evidence of strategic, in-depth thinking and analysis, especially related to social media principles
- Mechanics (such as grammar, spelling, punctuation)
- Organization of ideas and writing quality
- Conformance to AP format, style guides, standards covered in class, and assignment instructions
- Creativity
- Professional appearance

Sampling Workshop: Choosing Organizations

In class, students will discuss and vote on potential industries of study. Once the industry has been chosen, we will discuss the characteristics of a rigorous sampling process in order to choose an appropriate list of organizations to study.

Data Collection: Coding Workshops

In class, students will develop and learn to use a codebook. Students will learn to employ an iterative process to identify themes and patterns across the data. Students will practice "coding" together. Each student will be assigned a series of tweets to analyze. Then, using the codebook, each student will code the tweets.

Qualitative Data Analysis Workshops
Results: Identifying Themes and Patterns

In class, students will discuss the results. Students should come prepared by having read through their own codes, their classmates' codes, and writing a Discussion Board post. Together, students will identify the themes that emerge around these data. Students will learn how to write the results around a theme. Students will be assigned one theme/finding to write. You should write approximately 500 words. The goal of qualitative findings is an in-depth understanding of a phenomenon (as opposed to causation or generalization). You should:

- Providing rich descriptions.
- Use direct quotations so that each organization is represented in its own terms and ways of expressing itself.
- Keep quotations and descriptions in context.
- Identify any unique or unexpected patterns.

Qualitative Research Analysis

Writing the Results/Themes:

Format: *excellent assignments include:*

- Subheading(s)
- Direct Quotations
- Appropriate Citations

Content: *excellent assignments include:*

- Include relevant analysis of retweets, favorites, replies
- Ease in flow (i.e., organized and well-thought-out)
- Evidence (and accuracy) of findings
- Appropriateness of findings
- Data analysis, arguments, and/or findings that adequately answer the research question(s)
- A novel approach, proposing new ideas rather than just replicating existing research

Style: *excellent assignments include:*

- Concise, relevant, logical (use of past tense)
- Proper use of terminology and appropriate word choice
- Clear writing (e.g., free of jargon, cliché, or hyperbole)
- Sophisticated writing style (no passive voice, etc.)

Conclusions/Implications Discussion

In class, students combined all of the results into a comprehensive Results section. Students should come prepared by having read each other's themes. After discussing all of the results together, we will discuss the conclusions and implications of these results. In class, a guided brainstorming activity will be used to draw conclusions and implications. Based on the research findings, you will complete an in-class activity to develop a Top 10 List of evidence-based best practices to improve new media writing. The Discussion or Conclusion summarizes the results and suggests practical applications for those findings. Finally, we will discuss the research in its entirety. We will identify limitations and strengths of the research.

CHAPTER EIGHT

New Technologies FOR Social Media AND Public Relations Education

MELISSA D. DODD AND THIBAULT ADDA

Technology in education has enhanced students' learning environment and prepared them for their future professional careers. New technologies have been used to "accommodate student needs, promote student learning, and better prepare students for a digital society" (Zhao, 2007, p. 312). While classrooms of the past focused on overhead projectors (1950s), photocopiers (1970s), and VCRs, computers, and the Internet (1990s), the 21st century has been marked by the use of social media such as Facebook, Twitter, YouTube, and so on. New technologies such as tablets, smartphones, and wearables (e.g., smartwatches and Google Glass) in addition to evolving computers have facilitated the use of social media. The relationship between technology and social media is symbiotic. "New media" are operationalized and referred to in this chapter as the combination of social media platforms with the synergetic technologies that facilitate their use.

The new media continue to reshape the way we think, learn, work, and communicate. A nine-year longitudinal study by Wright and Hinson (2014) found that "various new, emerging and social media have brought dramatic changes to many aspects of public relations practice." Breakenridge (n.d.) posed that social media have revolutionized the way public relations professionals approach communication programs and campaigns. She states, "today, social media requires that we expand our KSAs [knowledge, skills, and abilities] for continued success in our public relations roles." This is perhaps evidenced by the emergence of social media dedicated courses for public relations majors, now a required course at many universities. For public relations students entering a competitive job market, understanding new media is a particularly important skillset. Indeed, Dishman (2014) argued that social media is the "new skills gap." She found that 90% of jobs in 2015 "require information and communication technology skills" with social media adding a projected $1.3 trillion to the U.S. economy. Many companies recruit Millennials to fill these positions because of the widespread assumption that they are fluent in new media, which leading social media educator Dr. William Ward described as "a mistake" (Dishman, 2014).

Research demonstrated that the inherent digital savvy of the Millennial generation is lacking in evidence (Hargittai, 2010), and systematic differences in abilities and the ways new media are used were present across groups of individuals. Students require training to effectively use new communication technologies and tools in the completion of tasks (Baan & Maznevski, 2008; Charsky et al., 2009; Rheingold, 2008), especially when it comes to using these tools as a strategic function of public relations (Anderson & Swenson, 2013; Gallicano, Ekachai, & Freberg, 2014; Miller Russell, 2007). Public relations educators of social media (and those who have integrated it as a major component in their courses) have struggled with issues of divide. Students do not enter courses with equivalent levels of new media experience, and public relations educators have been forced to bridge the gap between the most and least skilled students in their courses. A public relations instructor may easily have students enrolled in the same course who do not have accounts on major social networking sites (e.g., Facebook or Twitter) and lack relevant technological proficiencies, alongside students who maintain brand pages with communities of followers in the tens of thousands.

Public relations educators of social media perform several pedagogical functions in the classroom. Not only must these educators teach new media skills, but also they must be flexible enough in their teaching approach to inform both the least and most skilled students of new media. And, that is to say nothing of the actual teaching of public relations strategy as related to social media. So, the purpose of this chapter is (a) to explore the interdependent relationship between new communication technologies and social media, specifically as related to public relations education; (b) to discuss the issues of divide among students and pose how educators may best identify and bridge this gap for effective teaching of public relations and new media; and (c) to teach students how to strategically integrate new media into public relations campaigns.

LITERATURE REVIEW

An Operationalization

Many terms are used to refer to the new communication technologies, platforms, and tools that have evolved to be included in our understanding of social media. More often than not, terms are used interchangeably. Thus, it is important to begin with an operationalization of the key concepts used to distinguish these emergent platforms and tools. Safko (2012) defined *social media* simply as "the media we use to be social" (p. 3). Social media refer to platforms that allow users to connect, engage, and interact. Key examples boasting billions or multimillions of active users include Facebook, Google+, Instagram, Pinterest, Tumblr, Twitter, Snapchat, and YouTube. Kaplan and Haenlein (2010) explained, "Web 2.0 is a term that was first used in 2004 to describe a new way in which software developers and end-users started to utilize the World Wide Web" (pp. 60–61). They added that Web 2.0 platforms and technology define (and were defined by) social media (Kaplan & Haenlein, 2010).

The rise of new technologies and tools are inseparable from discussions of social media. For example, mobile technologies have facilitated the ease of access and frequency of use of social media. Mobile technologies have transformed the way we communicate and receive information both with and without respect to social media. Recent data suggested that for the first time in history, mobile Internet access has surpassed desktop computer access, and wearable technologies such as smartwatches and Google Glass were gaining mainstream attention (Lipsman, 2014; O'Toole, 2014; Perez, 2014). At present, wearables generally function in similar ways as mobile smartphone devices, allowing users to access the Internet, email and text messaging, mobile applications and social media separate from physical constraints. The major advantage of wearables is the convenience of access; that is, users are able to use the devices to a large extent hands-free. Analysts predicted that the convenience of wearables would eventually allow for the technologies to supplant "traditional mobile" (Kaplan & Haenlein, 2010).

Looking forward, the term *new media* may more rightly encompass both the social and technological aspects of the evolution of Web 2.0 in the classroom. According to Logan (2010), "new media refers to those digital media that are interactive, incorporate two-way communication, and involve some form of computing" (p. 4). Yet, the New Media Institute explains that the concept is amorphous, and thus, difficult to define.

> Loosely, new media is a way of organizing a cloud of technology, skills, and processes that change so quickly that it is impossible to fully define just what those tools and processes are. . . . Perhaps in searching for a suitable characterization for this network of tools and ideas is the idea of limitless possibility. Limitless possibility for communication, for innovation,

> and education is certainly a fundamental element that shapes our conceptions of new media usage. (Socha & Eber-Schmid, n.d.)

The State of New Media and Pedagogy

Although difficult to define, it is important for college educators to understand the prior experiences of incoming students and shortcomings of these experiences with respect to new media. In the United States, computer and Internet technology are widely accessible for educational purposes, yet there is a shortcoming in the implementation of these and other tools for instructional purposes. Prior experience predicted attitudes, abilities, and activities regarding new media (Ha & Shin, 2014; Hargittai, 2008, 2010).

Studies of new media use in higher education found that 100% of surveyed colleges and universities use social media for some purpose, with YouTube videos and blogs cited by faculty as the most frequently used for instructional purposes (Johnson, Adams Becker, Estrada, & Freeman, 2014). Faculty members generally believed that "the interactive nature of online and mobile technologies can create better learning environments" (Seaman & Tinti-Kane, 2013, p. 6); however, 65% reported that digital communication has increased the number of hours they work, and 48% said that digital communication has increased their levels of stress. All things considered, 41% of faculty indicated that they do use social media "frequently" as part of their teaching (Seaman & Tinti-Kane, 2013, p. 26).

According to Seaman and Tinti-Kane (2013), the most frequently used social media for teaching among higher education faculty were blogs/wikis, podcasts, and Twitter. Moreover, for blogs/wikis, faculty were more likely to require students to be content creators; with regard to podcasts, blogs, and Twitter, faculty were more likely to require students to be content consumers (read, watch, and listen). This seems to reflect current scholarly thought on the integration of new media into college curriculum where new media have been generally used to facilitate discussion in the classroom as well as serve as a reflection tool (Abe & Jordan, 2013; Elavsky, Mislan, & Elavsky, 2011).

New Media Users

A particularly important consideration for universities and educators was identifying and understanding the student users (and nonusers) of new media technologies and tools. Today, the typical college-aged student represents those most likely to use new media tools. The Millennial generation (18–29 years old) represents a generation that outnumbers their Baby Boomer parents by more than five million and who are defined by an Internet and media landscape that permeates nearly every facet of their daily lives. Prensky (2001) termed this generation "digital natives" with reference to the fact that Millennials are natives of the 21st-century digital language; non-Millennials or typical faculty members may be referred to as "digital immigrants" (Prensky, 2001; Bull et al., 2008). Millennials represent the heaviest users of new media with some 81% using social networking sites and 95% using the Internet (Madden, Lenhart, Duggan, Cortesi, & Gasser, 2013a; Madden et al., 2013b). Recent studies showed that 81% of college students use mobile technology such as laptops and smartphones (Belardi, 2015). However, Millennial users of the Internet and social media represent just the beginning of the new media revolution in education. In the next five to 10 years, universities and educators will see the widespread emergence of new technologies, namely in the form of wearables, in their classrooms. These new media technologies paired with the forthcoming Generation Z students will present a whole host of new challenges and opportunities.

Emergent Technologies

New media use and adoption change rapidly, but educators are not often early adopters. Until recently, desktop computers were the dominant method for accessing the Internet. However, the rapid growth in mobile technology has changed the landscape. Today, Internet access via mobile devices such as laptops, smartphones, and tablets is more frequent than access via desktop computers (Lipsman, 2014; O'Toole, 2014; Perez, 2014). For the first time in history, mobile usage accounted for 60% of all Internet consumption. The Pew Research Center predicts "by 2020, a mobile device will be the primary Internet connection tool for most people in the world" (Kaplan & Haenlein, 2010, p. 67).

More recently, the emergence of wearable technology has drawn widespread attention among individuals and educators alike. These devices perform the same functionalities as computers and smartphones but come in the form of glasses, watches, and bracelets. Wearable technologies work as a practical extension of the human body; that is, they augment users' reality. Augmented reality refers generally to a computer-mediated altering of users' real-world environment and experiences.

The implications of wearable technology and augmented reality in the classroom are many. For example, augmented reality allows for simulations and training exercises that may be valuable for public relations educators in the form of a crisis communication scenario. There is also the potential for holding class via augmented realities where the students and instructor are not physically located in the same place. However, an important consideration that may come sooner than later for educators is the potential for violations of the instructor's policies, dishonesty, or cheating. Already students may audio record instructors via smartphones without their knowledge, and wearables may allow for video recording with little capability for the instructor to recognize they are being recorded. Further, the ability and temptation to access Internet content will be great with imperceptible wearable devices in the form of glasses or watches, for example.

Issues of Divide

More than ever before, educators are seeking to harness the current state of new media for pedagogical effectiveness. Although Millennials represent "digital natives," universities and educators should be cautious about generalizing about students' abilities. Hargittai (2010) posed that despite widespread use, Millennials' digital savvy was not grounded in evidence. Further complicating the matter, research found systematic differences in how individuals use the Internet in their everyday lives with respect to access, gender, race/ethnicity, socioeconomic status, and experience (Hargittai, 2004, 2010). Hargittai (2010) explains, "internet know-how is not randomly distributed among the population, rather, higher levels of parental education, being a male, and being white or Asian American are associated with higher levels of Web-use skill" (p. 92). These variables are often inherent considerations for curriculum development, yet new media skills are often considered standard among Millennials in today's college classrooms.

Colleges and universities may better focus overall public relations curriculum development with empirical-based considerations about their students' understanding of new media technology and platforms. For example, community colleges, Historically Black Colleges and Universities, Hispanic-serving institutions, or those serving large numbers of nontraditional students may particularly benefit from a reexamination of student expectations surrounding new media. Accordingly, educators may better employ a valuable approach to technology-based course content and activities with an understanding of the digital divides that exist within their classrooms. Of particular relevance, research has found significant differences in academic performance as related to technology use (Jackson et al., 2008). Moreover, it is worthwhile for future communication professionals to understand the audiences they (or their companies) would engage with in social media spaces.

Hargittai (2010) stated, "people who have grown up with digital media are often assumed to be universally savvy with information and communication technologies. Such assumptions are rarely grounded in empirical evidence" (p. 92). Even when controlling for a host of other variables, however, people differed in their online abilities and activities (Hargittai, 2010). Hargittai (2008) similarly found that individuals with more usage experience were more likely to be users of new media. "Unequal participation based on user background suggests that differential adoption of such services may be contributing to digital inequality" (p. 276).

Further, differences in how users expect for new media to be used are also apparent in the literature. For example, Ha and Shin (2014) found that although students felt favorably toward social networking sites, they were still cynical about its use as a method of learning. This slight aversion was found to be the result of older students who were less willing to engage in educational interaction with the use of new media. Younger students did not harbor the same aversion, which was suspected to be due to familiarity of past experience with technologies.

METHOD

This chapter used an applied case to demonstrate how findings of a pretest and posttest measure may be implemented in the public relations classroom to bridge the new media skills gap. This approach allows for public relations educators to identify and evaluate new media skills as well as public relations strategy as applied to social media. The course employed for examination in this case consisted of 150 undergraduate advertising-public relations major students at the junior level or higher. The course content focused specifically on social media and public relations. Advertising-public relations students were given a pretest and posttest measure to assess issues of divide in the classroom. At the conclusion of the course, the issues of divide measure were administered again. The posttest administration allowed the instructor to compare results to the pretest to determine if abilities had improved between both the lowest- and highest-ability students.

Advertising-public relations students were also assigned a mid-semester team project. Students of varying new media ability levels were placed into groups and tasked with analyzing an organization's social media presence. Teams were then assigned to analyze and/or propose the integration of two social media sites and one mobile application into a campaign for the organization. The instructor was flexible as to which organizations were selected and, as a result, whether students focused on analyzing existing social media being used or proposing the integration of new social media. The purpose of the assignment was to further bridge the new media abilities gap both through collaborative engagement with classmates holding various ability levels and exposure to multiple new media. The assignment further sought to make clear the strategic implications of new media for public relations campaigns. Siemens (2005) proposed connectivism as a learning theory for the Digital Age. The theory emphasized how new media have created opportunities for learning and sharing information. Siemens (2005) explained that behaviorism, cognitivism, and constructivism were developed during a time when learning was not affected by technology. Connectivism emphasized the importance of learning as a process of connections to information (as opposed to the knowing of information). Downes (2007) explained, "at its heart, connectivism is the thesis that knowledge is distributed across a network of connections, and therefore that learning consists of the ability to construct and traverse those networks." In other words, pedagogical approaches using connectivism would facilitate the identification and navigation of effective networks. Connectivism poses that new media facilitates networks of nodes and connections wherein knowledge is kept, as opposed to within each individual. Thus, the process of identifying and navigating these networks is most salient.

For the social media course under analysis, the instructor first implemented quantitative pretest and posttest assessments to identify significant "issues of divide" among students enrolled in the course. Student teams in the course were also tasked with proposing how new media could be strategically integrated into public relations campaigns.

APPLICATIONS

Public relations scholars have advocated that new media be used in the classroom because of the prevalence of these tools in the industry (Miller Russell, 2007; Waters & Bortree, 2011; Waters & Robinson, 2008). Although Millennial students may represent those most familiar with new media technologies and tools, there have been systematic differences among groups regarding their attitudes, abilities, and activities of use. Further complicating the integration of new media into public relations education is the fact that even when Millennials do know how to use the technology, they need training to learn how to use it as a strategic public relations function.

Pretest and Posttest Assessment

Early in the course, students were asked to indicate the extent of their familiarity with specific social media platforms and new technologies. Students were also asked to evaluate their perceptions of self-efficacy and identify the ways they currently use new media. A pretest measure was administered online at the beginning of the course (see Assignment 8.1). Student data were aggregated so as not to reveal any identifying information, and results were presented in person during the following class meeting. The instructor then led students in discussions about issues of divide based on results of the pretest measure. The instructor asked students to break into small discussion groups where they shared their different ability levels with regard to new media. This created an initial understanding of the different skill levels among classmates. Ideally, for the less-skilled users, the presentation of the assessment results demonstrated their job-market "competition" for public relations positions that emphasized social media skillsets. Likewise, for the more-skilled users, the presentation helped encourage a tolerance and understanding of skills-based new media course content, with which they may already have familiarity. For example, the student who already managed brand pages with tens of thousands of followers would be less likely to need basic information about how to use Twitter. However, the skilled users were much more understanding of such course content, and even willing to help classmates reach learning objectives, when they realized how advanced their skillsets were. Indeed, Elavsky et al. (2011) similarly found that the use of new media in a large-lecture course generally resulted in positive outcomes overall. They state that

> Twitter dramatically altered the in-class communication patterns, reconfiguring the discourse therein into multiplatform engagements by which online/offline utterances intermingled, stimulating new trajectories for deliberation that significantly expanded the scope and potential of the class discussions beyond traditional contextual parameters. Fostered almost singularly by student input, the Twitter stream discourse deepened and extended the class potential for engagement with the course themes in novel ways (p. 225).

The implementation of a pretest assessment identified potential issues of divide, and the presentation of the results of assessment fostered mutual understanding and a collaborative learning environment.

Further, the instructor was able to focus efforts accordingly for both the most- and least-skilled students, tailoring the class to bridge the skills gap. Outside of class, the instructor divided students into two relative groups based on assessments of their abilities: low- and high-new media skills. Throughout the duration of the course, the instructor was able to address generally the different needs of both groups. For example, in discussing Twitter, the instructor asked a high-skill student to demonstrate the basic how-to information during the class meeting. The remainder of the class was dedicated to discussing Twitter as a tool for public relations, presenting case study examples and best practices. The instructor would make reference to the different groups with regard to the emphasis they should put on activities and assignments. The high-skill group, who already were active on Twitter, were tasked with focusing on their personal branding and message strategies for the course, while the low-skill group was encouraged to "get their hands dirty" by live tweeting an event and following industry influencers. To be clear, there was no demonstrable difference in the assignments given to students; however, the students were often reminded that the onus to develop a successful online personal brand was on them.

Also during the class meeting where pretest assessment results were presented, the instructor broke students into small groups (four to six students each) and asked them to discuss their individual differences with regard to new media uses, attitudes, and abilities. Students discussed their diverse backgrounds and experiences. Issues of divide that were assessed and presented quantitatively to students were now afforded context. For example, students realized that not all classmates had equivalent experiences or backgrounds. Not all students had laptops that were advanced enough to bring to class because they were too heavy or had low battery life. Some students took notes on their mobile phones, while others did not own a smartphone or preferred to avoid the distractions of new media and took notes with paper and pen.

Students' attitudes toward new media also differed. Students without accounts on major sites like Facebook or Twitter had privacy concerns or just didn't see a need for it in their personal lives. Students addressed their comfort levels with communicating with instructors and classmates via new media. Discussions further gave rise to salient issues of privacy and the mixing of the personal with professional. The new media have redefined the student-instructor relationship, while allowing for new communication channels to communicate course content. For example, Waters and Bortree (2011) found that public relations students with higher levels of communication apprehension were less likely to communicate with their instructors outside the classroom, but more likely to communicate with instructors via social networking sites with greater frequency than those students with low levels of communication apprehension.

McCorkindale (2013) detailed that public relations educators were increasingly integrating new media into their courses and connecting with their students to demonstrate emerging communication strategies. However, she states, "balancing personal and professional use is important because instructors may represent best practices for their students." Privacy for both students and instructors remained a major concern for the integration of new media into the classroom (Lewis & West, 2009; McCorkindale, 2013). Many educators and students remain hesitant about blurring the lines between their personal and professional lives, so addressing these issues at the onset of the course has been essential for mutual understanding and respect for others. Again, this also brought to light for students the various issues organizational stakeholders would perceive in engaging with their company (or clients' companies) via social media.

Finally, at the end of the semester, the instructor asked students to again participate in the quantitative assessment. Using the data, the instructor was able to determine if students of different levels had improved in the areas most relevant to them. For example, the students who were already at a relatively high level of new media technical skills were able to improve on communication strategy, campaigns, analytics, evaluation, etc., in a new media context. Likewise, students with low levels of new media technical skill were able to improve on understanding the actual platforms, tools, and technologies. For the purposes of data analysis, low- and high-skilled students can even be separated into groups for the performance of statistical difference tests.

The implications of this approach are two-fold. First, instructors are able to refocus efforts where they notice deficits in future course offerings. Public relations educators should seek to address the appropriate mix of new media technological and technical skills while addressing appropriate public relations strategy as related to social media. Instructors should also be flexible enough in their teaching approach to inform both the least and most skilled students of new media.

New Media Campaign Integration

In addition to ongoing assessment, the instructor sought to further bridge new media ability gaps through a team-based assignment where students of varying ability levels collaborated on a public relations campaign proposal. This assignment simultaneously focused on the strategic role of new media for public relations, reinforcing the idea that new media use for public relations should move beyond the tactical level to connect with overarching organizational goals and objectives (see Assignment 8.2).

Students were first divided into teams of five. The instructor divided students of various skill levels evenly among teams. The instructor asked that the students give their newly formed public relations firm (i.e., their teams) a name. This created a greater sense of shared purpose and cohesion, despite disparate ability levels. Students were then asked to select an organization (on a first-come, first-serve basis approved by the instructor). Organizations could be large or small; local, national, or international; corporate, nonprofit, or governmental. Student teams were then tasked with (1) analyzing the organization's existing use of social media and (2) strategizing and proposing the integration of new media for the organization to communicate with a specified audience. The culmination of this project resulted in a team research paper and Prezi presentation. Student teams presented their work to their classmates, which allowed for exposure to various new media in a strategic public relations context.

This assignment would best be executed as a mid-semester or final course project. This would allow for a pretest assessment and discussion of abilities to have already occurred as well as new media course content that contributes to an overall understanding of the strategic role of new media for public relations campaigns.

DISCUSSION

This chapter conceptualized "new media" as the intersection of social media platforms and the technology that facilitates them. For public relations education, this has been a particularly salient issue that was best addressed through the pulling together of existing secondary literature to provide a comprehensive description of the new media user and issues of divide that arose in the classroom. Using a pretest and posttest assessment, discussion exercises, and a new media campaign integration assignment, students quickly realized that their classmates come from diverse backgrounds and have a variety of attitudes, abilities, and uses with regard to new media. This fostered an environment of mutual understanding and collaboration from the onset of the course.

Beyond facilitating a better learning environment, public relations students need to be well-versed in new media as part of their future professions. Understanding the issues of divide among their organizational stakeholders facilitates more effective communication programs and campaigns. The new media campaign integration assignment reintroduced the idea of understanding stakeholder uses of new media as part of an overarching campaign strategy. Moreover, public relations professionals are considered the "corporate conscience" or ethical counselor to management for their organizations (Bowen, 2008; Ryan & Martinson, 1983). Thus, it is valuable to task students with discussing individual differences that give way to conversations about diversity of attitudes, abilities, and uses as well as issues of privacy.

It is also valuable for the public relations educator to identify and evaluate issues of divide. By realizing the differences in attitudes, abilities, and uses among students, the instructor may tailor the course to bridge the new media skills gap. The instructor should construct course content and activities according to student skill levels, both with regard to new media technical and technological skills and public relations strategy using new media. Students should be familiar with the latest technological devices as well as social media tools, and instructors may use the pretest and posttest assessment to continually evaluate effectiveness in their courses.

Although the assessment is not unique, the addition of technology and issues of divide to social media knowledge and skills paints a more comprehensive picture of students' actual abilities. Public relations educators may also engage students in the following supplemental course assignments: (1) the identification and discussion of traditional and new media theories; (2) a class discussion about how issues of divide affect communicating with different audiences; and (3) the required use of new media.

The identification of traditional and new media may facilitate the connection of theory to practice. Traditional theories may be applied to a new media context (e.g., media richness theory) and compared to new media-based theories (e.g., connectivism) for a discussion of the public relations implications.

The issues of divide are important considerations both for educators and public relations professionals who may seek to communicate with audiences. A class discussion of diverse stakeholders allows students to identify characteristics among diverse publics with whom they seek to communicate. This brings to light the importance of strategy in connecting new media to the audience. For example, using YouTube to communicate with the 65 and older demographic might not be appropriate.

The required use of new media is recommended. For example, not only does maintaining a professional blog and Twitter account allow students to produce portfolio samples and connect with industry professionals around the world, but it also exposes them to the new media tools used by public relations professionals today. This facilitates training and experience using these tools as professionals before they enter the increasingly competitive job market. Moreover, comprehensive social media management training exists at no cost for educators and their students (e.g., HootSuite University). Instructors may integrate such training and certification tools into coursework, further bridging the new media skills gap both within and outside the classroom.

REFERENCES

Abe, P., & Jordan, N. A. (2013). Integrating social media into the classroom curriculum. *About Campus, 18*(1), 16–20. doi:10.1002/abc.21107

Anderson, B., & Swenson, R. (2013). What should we be teaching our students about digital PR? Collaborating with top industry bloggers and PR Twitter chat professionals. *Teaching Public Relations, 87*. Retrieved from http://www.aejmc.us/PR/tpr/TPR87fa13.pdf

Baan, A., & Maznevski, M. (2008). Training for virtual collaboration. In J. Nemiro, M. Beyerlein, L. Bradley, & S. Beyerlein (Eds.), *The handbook on high-performance virtual teams* (pp. 345–365). San Francisco, CA: Jossey-Bass.

Belardi, B. (2015, March 9). *Report: New McGraw-Hill education research finds more than 80 percent of students use mobile technology to study*. Retrieved from http://www.mheducation.com/about/news-room/report-new-mcgraw-hill-education-research-finds-more-80-percent-students-use-mobile

Bowen, S. A. (2008). A state of neglect: Public relations as "corporate conscience" or ethics counsel. *Journal of Public Relations Research, 20*, 271–278.

Breakenridge, D. (n.d.). Social media knowledge, skills, and abilities. *Public Relations Society of America*. Retrieved from http://www.prsa.org/jobcenter/career_resources/resource_type/tools_tactics/social_media

Bull, G., Thompson, A., Searson, M., Garofolo, J., Park, J., Young, C., & Lee, J. (2008). Connecting informal and formal learning: Experiences in the age of participatory media. *Contemporary Issues in Technology and Teacher Education, 8*(2), 100–107. Retrieved from http://www.citejournal.org/articles/v8i2editorial1.pdf

Charsky, D., Kish, M. L., Briskin, J., Hathaway, S., Walsh, K., & Barajas, N. (2009). Millennials need training too: Using communication technology to facilitate teamwork. *Techtrends: Linking Research and Practice to Improve Learning, 53*(6), 42–48. doi:10.1007/s11528-009-0342-2

Dishman, L. (2014, September 16). The job skills gap you haven't considered. *Fast Company*. Retrieved from http://www.fastcompany.com/3035669/the-future-of-work/the-job-skills-gap-you-havent-considered

Downes, S. (2007, February 3). *What connectivism is*. Retrieved from http://halfanhour.blogspot.com/2007/02/what-connectivism-is.html

Elavsky, M., Mislan, C., & Elavsky, S. (2011). When talking less is more: Exploring outcomes of Twitter usage in the large-lecture hall. *Learning and Media Technology, 36*(3), 215–233. doi:10.1080/17439884.2010.549828

Gallicano, T. D., Ekachai, G., & Freberg, K. (2014). The infographics assignment: A qualitative study of students' and professionals' perspectives. *Public Relations Journal, 8*(4). Retrieved from http://www.prsa.org/Intelligence/PRJournal/Documents/2014GallicanoEkachaiFreberg.pdf

Ha, J., & Shin, D. H. (2014). Facebook in a standard college class: An alternative conduit for promoting teacher-student interaction. *American Communication Journal, 16*(1), 36–52. Retrieved from http://ac-journal.org/journal/2014–2015/Vol16/Iss1/ACJ_2014-015_Shin3.pdf

Hargittai, E. (2004). Internet access and use in context. *New Media and Society, 6*(1), 137–143. doi:10.1177/1461444804042310

Hargittai, E. (2008). Whose space? Differences among users and non-users of social network sites. *Journal of Computer-Mediated Communication, 13*(1), 276–297. doi:10.1111/j.1083-6101.2007.00396.x

Hargittai, E. (2010). Digital na(t)ives? Variations in Internet skills and uses among members of the "net generation." *Sociological Inquiry, 80*(1), 92–113. doi:10.1111/j.1475-682X.2009.00317.x

Jackson, L. A., Zhao, Y., Kolenic III, A., Fitzgerald, H. E., Harold, R., & Von Eye, A. (2008). Race, gender, and information technology use: The new digital divide. *Cyberpsychology & Behavior, 11*(4), 437–442. doi:10.1089/cpb.2007.0157

Johnson, L., Adams Becker, S., Estrada, V., & Freeman, A. (2014). *NMC Horizon Report: 2014 Higher Education Edition*. Austin, TX: The New Media Consortium.

Kaplan, A. M., & Haenlein, M. (2010). Users of the world, unite! The challenges and opportunities of social media. *Business Horizons, 53*(1), 59–68. doi:10.1016/j.bushor.2009.09.003

Lewis, J., & West, A. (2009). "Friending": A London-based undergraduate's experience of Facebook. *New Media & Society, 11*(7), 1209–1229. doi:10.1177/1461444809342058

Lipsman, A. (2014, July 11). *Is mobile bringing about the death of the PC? Not exactly*. Retrieved from http://www.comscore.com/Insights/Blog/Is-Mobile-Bringing-About-the-Death-of-the-PC-Not-Exactly

Logan, R. K. (2010). *Understanding new media: Extending Marshall McLuhan*. New York, NY: Peter Lang.

Madden, M., Lenhart, A., Cortesi, S., Gasser, U., Duggan, M., Smith, A., & Beaton, M. (2013b, May 21). Teens, social media, and privacy. Retrieved from http://www.pewinternet.org/files/2013/05/PIP_TeensSocialMediaandPrivacy_PDF.pdf

Madden, M., Lenhart, A., Duggan, M., Cortesi, S., & Gasser, U. (2013a, March 13). Teens and technology 2013. Retrieved from http://www.pewinternet.org/files/old-media//Files/Reports/2013/PIP_TeensandTechnology2013.pdf

McCorkindale, T. (2013). Will you be my friend? How public relations professors engage with students on social networking sites. *Teaching Public Relations, 85*. Retrieved from http://www.aejmc.us/PR/tpr/TPR85sp13.pdf

Miller Russell, K. (2007). Using weblogs in public relations education. *Teaching Public Relations, 73*. Retrieved from http://www.aejmc.us/PR/tpr/73MillerRussell.pdf

O'Toole, J. (2014, February 28). Mobile apps overtake PC Internet usage in U.S. *CNN Money*. Retrieved from http://money.cnn.com/2014/02/28/technology/mobile/mobile-apps-internet/?iid=EL#TO

Perez, S. (2014, August 21). Majority of digital media consumption now takes place in mobile apps. *Techcrunch*. Retrieved from http://techcrunch.com/2014/08/21/majority-of-digital-media-consumption-now-takes-place-in-mobile-apps/

Prensky, M. (2001). Digital natives, digital immigrants. *On the Horizon, 9*(5), 1–6. Retrieved from http://www.marcprensky.com/writing/Prensky%20-%20Digital%20Natives,%20Digital%20Immigrants%20-%20Part1.pdf

Rheingold, H. (2008). Using social media to teach social media. *New England Journal of Higher Education, 23*(1), 25–26. Retrieved from http://files.eric.ed.gov/fulltext/EJ850702.pdf

Ryan, M., & Martinson, D. L. (1983). The PR officer as corporate conscience. *Public Relations Quarterly, 28*(2), 20–23.

Safko, L. (2012). *The social media bible: Tactics, tools & strategies for business success* (3rd ed.). Hoboken, NJ: John Wiley & Sons.

Seaman, J., & Tinti-Kane, H. (2013). *Social media for teaching and learning.* Retrieved from http://www.pearsonlearningsolutions.com/assets/downloads/reports/social-media-for-teaching-and-learning-2013-report.pdf#view=FitH,0

Siemens, G. (2005). Connectivism: A learning theory for the digital age. *International Journal of Instructional Technology and Distance Learning, 2*(1), 3–10. Retrieved from http://www.itdl.org/journal/jan_05/article01.htm

Socha, B., & Eber-Schmid, B. (n.d.). *What is new media?* Retrieved from http://www.newmedia.org/what-is-new-media.html

Waters, R. D., & Bortree, D. S. (2011). Communication with the millennials: Exploring the impact of new media on out-of-class communication in public relations education. *Teaching Public Relations, 80.* Retrieved from http://www.aejmc.us/PR/tpr/TPR80sp11.pdf

Waters, R. D., & Robinson, J. (2008). Blogging 101. Introducing blog management into public relations curriculum. *Teaching Public Relations, 74.* Retrieved from http://www.aejmc.us/PR/tpr/74waters.pdf

Wright, D. K., & Hinson, M. D. (2014). An updated examination of social and emerging media use in public relations practice: A longitudinal analysis between 2006 and 2014. *Public Relations Journal, 8*(2). Retrieved from http://www.prsa.org/intelligence/prjournal/documents/2014wrighthinson.pdf

Zhao, Y. (2007). Social studies teachers' perspectives of technology integration. *Journal of Technology and Teacher Education, 15*(3), 311–333.

ASSIGNMENT 8.1. PRETEST AND POSTTEST ASSESSMENT

Please respond with your agreement to the following items from Strongly Disagree (1) to Strongly Agree (5).

In general, I would consider myself to have a strong understanding of the following social media in regards to being both (1) an observer/reader of the media and (2) a creator or maintaining/posting to the media.

1. Analytics Services (Klout, HootSuite, Google/Facebook/Twitter Analytics, Web Traffic Analyzers)
2. Audio Creation/Sharing Sites (Audacity, iTunes)
3. Blogs (Tumblr, WordPress, Blogger)
4. Forums or Message Boards (Reddit)
5. Geo-location (Foursquare)
6. Microblogs (Twitter)
7. Photo Sharing Sites (Pinterest, Photobucket, Flickr)
8. RSS, Feed Readers (Really Simple Syndication)
9. Search Engines (Google, Bing, Stumble Upon, Facebook Graph Search)
10. Social Networking Sites for Personal Use (Facebook)
11. Social Networking Sites for Professional Use (LinkedIn)
12. Social News Sites (Flipboard, Zite)
13. Social Bookmarking Sites (Delicious, Digg)
14. Video Creation/Sharing Sites (YouTube, Vimeo)
15. Virtual Worlds and Gaming (SecondLife, World of Warcraft)
16. Wikis (Wikipedia)

In general, I would consider myself to have a strong understanding of the following technology:

1. Internet
2. Laptop
3. Smartphone
4. Tablet
5. Wearable technology

In general, I perform the following activities on the Internet:

1. Informational use/research, banking, reading, searching for jobs
2. Recreational use/entertainment, games, gambling, fun
3. Communication for professional purposes
4. Communication for personal purposes
5. Other (please specify)

In general, I understand how new media are used for the following public relations activities:

1. Campaigns
2. Strategy
3. Research
4. Evaluation

Demographic Information

1. Gender
2. Race/Ethnicity
3. Parents' highest education completed
4. Average hours/week spent on the Internet

ASSIGNMENT 8.2. NEW MEDIA CAMPAIGN INTEGRATION PROJECT

Overview:

In groups of five, students will select an organization (on a first-come, first-serve basis and approved by the instructor). Organizations may be large or small; local, national, or international; corporate, nonprofit, or governmental. Students must then (1) analyze the organization's existing use of social media and (2) strategize and propose the integration of new media for the organization to communicate with a specified audience.

This project must analyze and/or integrate a combination of at least two social media sites and one mobile application. A five-page paper detailing the aforementioned criteria is required as well as a group Prezi presentation. Note: This Prezi makes a great addition to your online portfolio, and can be posted to your LinkedIn profile.

Research Paper:

Student teams are required to provide brief introductory information surrounding the organization, itself, and conduct extensive research on the current organizational use of new media. The research paper should provide adequate background and introductory information so that an unfamiliar reader would get a sense of the organization and its current new media efforts. Students may answer questions such as the following:

- What social media platforms/apps is the organization using?
- With whom are they seeking to communicate using these media?
- What kinds of content are they posting?
- How is the organization effectively using new media?
- How is it ineffectively using new media?
- Which new media are the organization *not* using?

Next, based on the concepts and skills learned in this course, students should strategize and propose the integration of new media for the organization they've selected and aimed at communicating with a specified audience. A good starting point for this is to consider the goals of the organization and their existing new media use. Now, consider how that organization could achieve those goals using their existing new media more effectively and/or using additional social media tools. Consider the unique wants, interests, and needs (WINs) of the audience you've identified.

- How can we use new media to achieve the organizational goal(s) while addressing the audience's WINs?
- Are there additional social media tools that the organization should be using?
- Or, should your strategy be more related to improving the current use of new media tools (for example, improved content)?

You might set goals for the campaign and consider how those would be evaluated for effectiveness too. In what ways does your proposed campaign enhance the organization's current new media use, strategy, or overall goals? And so on.

The combination of the existing social media analysis and proposed campaign integration should result in a research paper that addresses a combined total of at least two social media sites and one mobile application. Otherwise, be creative and have fun with your campaign proposal! For any of you that hope to work as professional communicators in the social media world, you will likely work as part of an agency that pitches campaign ideas to potential clients. Only the most creative and strategic ideas get the client's business.

Prezi Presentation:

As a result of extensive background research and collaborative efforts aimed at a creative, strategic campaign proposal on behalf of their organization, students will produce a Prezi presentation (www.Prezi.com). The focus of your Prezi is to (1) give an overview of the organization and its current social media use and (2) present your team's proposed new media integration campaign.

The information presented in your Prezi should result in a five- to seven-minute presentation. Whenever appropriate, the information presented should be supported by research and citations you've found as part of your research (using APA format).

Your team should also consider the following: color scheme; use of appropriate fonts; use of graphics and visuals; and strategic use of videos, charts, and graphs. Be sure everyone in your team has completed the Prezi tutorial.

You will send the instructor a web link to your Prezi set to public view or exported in PDF format. Also note that you can download Prezi presentations to your desktop. For the presentation portion of this project, you'll need to be sure that each team member has access to a copy of the presentation.

PART THREE

Applying Social Media in Teaching Journalism

CHAPTER NINE

Challenging THE Newsroom Paradigm

Four Nations' Journalism Students Interrogate Global Issues Through Social Media

DAVID BAINES AND MELISSA WALL

Social media have helped to disrupt, dislocate, and in some cases dissolve many boundaries that have long characterized journalism practice: boundaries between the professional and amateur; local and global; producer and consumer; process and product; mass and interactive communication. Yet, journalism education does not seem to have responded fast enough or dramatically enough to these enormous changes. One reason has been that educators have oftentimes looked toward the journalism industry for ways to adapt, even though critics believe traditional journalism has failed to adequately respond to the scope of change (Anderson, 2013; Stencel, Adair, & Kamalakanthan, 2014). This follows a historic pattern in which journalism educators preparing students for careers in the field often modeled their campus-based teaching and learning facilities on the routines and practices of the industry. They valorized the newsroom as a teaching and learning space where the classroom experience was as close to work experience as they could make it. The intention was to replicate the professional experience, creating a dress-rehearsal for a career in the industry. But has this been the best way to approach the dramatic changes rocking the field?

This chapter suggests not, focusing on a collaborative global project, the Pop-Up Newsroom, which challenged these tendencies and served as a model for ways in which social media could help educators rethink journalism education. This study joins a growing group of journalism educators who have called for community-based, experimental approaches to journalism education (Baines & Li, 2013; Berger, 2011; Flew & Wilson, 2010; Robinson, 2013). Critics such as

Mensing (2010) have argued that journalism educators take advantage of the university environment to innovate and imagine new models, suggesting they move their focus away from "the transmission-driven, industry-conceived model of journalism" and toward the wider community (p. 512). It therefore behooves journalism educators to encourage students to question the values and practices that have grown out of the conventional newsroom, and the boundaries it has erected, and explore new ways to practice journalism. It is also up to educators to steer students toward explorations of the way tools such as social and mobile media can be deployed to reimagine journalism.

The purpose of this chapter is thus threefold: (a) to offer a critical reflection, informed by Castells' concepts of the network society and mass self-communication, on the manner in which emergent forms of social media platforms and digital devices are disrupting and reshaping practices and values embedded in the industry-conceived model of journalism; (b) to describe an innovative project in which journalism students from several countries adopted a range of such platforms and devices to report on the critical issue of poverty across national and cultural boundaries; and (c) to interrogate the manner in which those students internalized this experience to challenge their understandings of established professional practices and values, and to reflect on alternative, networked approaches to journalism. This chapter maps a course beyond familiar newsroom boundaries that journalism educators might offer to their students to help them to explore the constantly morphing frontiers of journalism.

LITERATURE REVIEW

Research on journalism structures and processes has repeatedly shown that the newsroom was a place of socialization and thus control over journalism practices and values (Domingo, 2008; Gans, 1979). This control was manifested at the level of the journalists in the values they were expected to learn and enact through the routinization of news production and at the broader ideological level through the influence of ownership models, which often interlocked with political elites (Gans, 1979). Hierarchically organized newsrooms often produced journalism that legitimized the status quo, drawing on the voices of elite sources and excluding those at the margins.

Such organizational settings have given rise to newsroom cultures that foster group-centric attitudes (Tuchman, 1978). Thus, newsroom walls were not only physical boundaries but also marked symbolic divisions between producers and consumers of news. Traditionally, those within decided what those without would see and hear of their world, but those without were not privy to how those decisions were made, nor how the news was produced. Newsroom boundaries also gave structure to journalism work. They provided journalists with clear criteria for evaluating their performance, for example, the delivery of a well-done story in the judgment of peers. By providing routines and organizational hierarchy, they created a sense of security. Some researchers argue this has led to "a deeply self-referential professional culture," which often guarded its own values and its own traditional workspace (Deuze, 2008, p. 857). In fact, much of the boundary work literature within journalism studies has shown the ways professional journalists have long sought to draw lines between themselves and others, including more recent threats to their authority such as citizen journalists (Holton, Coddington, & Gil de Zúñiga, 2013).

However, some researchers argued that these tendencies have been evolving as today's newsrooms undergo profound changes created by social, technological, and economic shifts, and journalists must come to terms with a host of changes including the spread of participatory cultures and heightened emphasis on interactivity (Anderson, Bell & Shirky, 2012; Robinson, 2011; Usher, 2014). Anderson (2013) documented the rise of "insurgent" newsrooms operating on different models from the legacy news media, opening their newsrooms to volunteers, institutions, and others sharing news production processes. In sum, the long-standing control enacted by professional journalism is in competition today with a desire by many for openness (Lewis, 2012).

Networked Journalism and Mass Self-Communication

In this project, we considered these changes through the application of two concepts explicated by Castells (1996, 2007), which provided a framework for understanding the dramatic reshaping of journalism: the network society and mass self-communication. The network society is a new structure or logic for society that orders human experience (Van Dijk, 2012). It can be thought of as a web-like form generating more widespread and faster connections between nodes than ever before. This communication structure created a "space of flows," which indicated its fluidity but did not mean that all the nodes within the network were equal or that social hierarchies and inequities disappeared (Van Dijk, 2012). Even in a network society, power can and does accumulate. That said, this power can be challenged from within by using its own logic. Castells suggested this occurred when the experiences of local communities were connected globally through networked communication. Related to these phenomena is the rise of what Castells (2007) called "mass self-communication," which was "the emergence of a new form of communication related to the culture and technology of the network society, and based on horizontal networks of communication" (p. 239). Using new digital tools, individuals could potentially connect with others who were globally distant and grassroots organizations could challenge existing hierarchies.

Bardoel and Deuze (2001) described networked journalism as converged media, or the process by which the various technological forms (print, video, audio) could be combined when they were digitized. But networked journalism is more than the merging of forms. As Beckett and Mansell (2008) noted, networked journalism created a new news logic that challenged the historical trend of one-way news delivery to passive audiences. In doing so, news potentially has become more diverse and pluralistic, opening spaces for other, nontraditional voices such as when professional journalists are joined by a range of others from citizen journalists to bloggers in a global network of news production (Heinrich, 2011).

Thus, networked journalism enabled content to be distributed much more widely than ever before through potentially global information flows in which both citizens and professionals were producers. Indeed, it contributed to the dissolution of traditional understandings of the audience and how professional news producers might interact with nonprofessionals, as amateurs increasingly contributed content to professional news outlets. The journalist became only one "node in a network that functions to collect, process, and distribute information" (Van der Haak, Parks, & Castells, 2012, p. 2927).

Geographical boundaries for the news itself were further challenged as once-local publications or broadcasters, closely bounded geographically and concerned with local affairs, became part of national or global media conversations. The communities with which they were concerned may also have developed global as well as local concerns and connections (Deuze, 2008).

The impact of these changes has been significant. For example, networked journalism might contribute to the expansion of genres of storytelling traditionally used by journalism, and thus might create new forms of news. Long-accepted practices such as the idea that a story is reported and edited in a newsroom and then delivered to an audience in a linear manner has been transformed (Karlsson & Strömbäck, 2010). Stories are reported in fragments—perhaps through Twitter or other social media posts—as they happen. These fragments may then be further changed as the reporting continues and numerous versions are posted on a news site as events unfold (Karlsson & Strömbäck, 2010). That said, some critics argued that such practices had degraded journalism, producing a lower quality of reporting and contributing to ethical problems (McNair, 2009). Despite these concerns, Van der Haak, Parks, and Castells (2012) called networked journalism "the future of journalism" (p. 2927).

These dramatic developments have led to questions about the efficacy of limiting reporting to traditional, spatially constrained newsrooms. Such spaces reflected what has been called an "industrial" model of the newsroom production processes that no longer entirely applies (Mensing, 2010). Heinrich (2011) noted that journalism was "moving away from traditional place-based modes of production and dissemination" (p. 8). Indeed, Beckett and Mansell (2008) argued that networked journalism allowed "new forms of boundary crossing" that "may provide a foundation for public dialogue that enables stories about distant others to be told" (p. 92). Journalism may be becoming more connected, diverse and engaged, potentially allowing journalists to overcome problems such as the entrenched othering of marginalized communities; yet, as Deuze (2007) reminded us, the bureaucratic structures remained in place in most mainstream newsrooms.

In contrast with these positive assessments of networked journalism, some researchers have warned that as more nonprofessional content became available online coupled with significant declines in resources in many Western news outlets, newsrooms decreased the amount of original reporting (Thurman & Myllylahti, 2009). In light of this, it was important to understand how journalism students responded to working without conventional professional structures and boundaries and whether (and how) they could adapt to practicing networked journalism with an emphasis on journalism as process rather than product, using social media platforms and mobile media devices in a media environment characterized by mass self-communication. So,

RQ1: How do journalism students negotiate practicing journalism as networked process rather than the traditional production of a linear, polished product?

RQ2: How do journalism students negotiate the absence of spatial and symbolic newsroom boundaries, which characterize industrial models of journalism?

METHOD

This section details a social media–based journalism project that the authors launched to better teach students about mobile media and their potential for transforming news and creating global connections.

Launching a Global Newsroom

This project initially grew out of a conversation between the authors at the 3rd World Journalism Education Congress held in Belgium in 2013. They had both created projects at their universities based on networked principles, employing the term *pop-up* to characterize what the students did (Wall, 2015), and decided to team up to host a transnational pop-up news project with student journalists from around the globe reporting on the same topic for 24 hours.

Partners. The first step was to identify potential partners. One obstacle was finding universities that were in session at the same time. Participating schools also had to be willing to immerse their students in reporting with social media tools. The team ultimately consisted of four universities: the Asian College of Journalism (ACJ) in India; California State University–Northridge (CSUN) in the United States; National Chung Cheng University in Taiwan (NCCU); and Newcastle University (NU) in the United Kingdom. Most communication was through email; a Google Hangout was held to discuss logistical details and get to better know each other.

Topic. Previous pop-ups at CSUN and NU had been focused mainly on events such as local festivals. For this project, there was no obvious common event for the countries to cover, so a topic area was selected. This too had to be broad enough to have relevance for all participants. The instructors wanted a serious topic to help students understand how social media could be used to report on significant issues and not just trivia. Thus, income inequality was first chosen. The United Kingdom initially focused on "austerity cuts" from the 2008 global recession that were still being felt in that country (they later changed to a wider perspective embracing "poverty and deprivation"); India focused on "deprivation" and poverty; the United States and Taiwan chose "poverty."

Student Participation. 127 journalism students from the United States, United Kingdom, and India took part in the project. The Taiwan team was part of a domestic project that took place earlier in the semester and included more than one school. Details about the numbers of their participants were incomplete and reflected the challenges in carrying out an international educational project.

With the exception of NCCU, all were graduate students. Other students, journalism instructors, professional journalists, and community members participated by retweeting coverage and, in the case of the United Kingdom, joining the pop-up news operation. Requirements varied by school. For the U.S. team, participation was mandatory through a multistage course assignment (see Assignment 9.1). For the UK team, participation was mandatory for students in the graduate program in International Multi-Media Journalism, but voluntary for students from other journalism and PR graduate programs. Participation was mandatory for the Indian students in the ACJ's Multimedia Journalism program. Neither India nor the United Kingdom graded the exercise. Students were encouraged to make connections using social media, particularly Facebook and Google Hangouts.

Covering the Story. Many of the students involved researched issues surrounding poverty, deprivation, and the media before the event in an effort to avoid overly simplifying the topic. For example, Newcastle students read *Reporting Poverty in the UK: A Practical Guide for Journalists* by the nonprofit Joseph Rowntree Foundation. They also heard a lecture by Professor of Education Technology Sugata Mitra, at Newcastle, whose work on education and poverty is internationally known.

CSUN students wrote "mini" literature reviews of academic research about their specific area of interest such as homeless veterans and read a special issue of the *Journal of Mass Media Ethics,* which ran a series of articles critiquing traditional news coverage of poverty. ACJ students took part in a seminar and workshop with Professor K. Nagaraj, an economist and poverty expert at the Madras Institute of Development, noted for his study of farmers in India driven by poverty to commit suicide. Videos of a series of interviews with Nagaraj were posted to YouTube so that any student participating in the project could view them; the readings done by the United States and the United Kingdom were also shared should students want to see what others were reading; a recording of Mitra was shared as well.

Common across countries, students found that accessing the poorest and most vulnerable generally meant contacting organizations that served those communities. For the Newcastle students, where subjects were not available on the day of the event, some students prepared content in advance for later release through social media. CSUN students identified sites or events where they could be physically based on the day of reporting; while pre-interviews were allowed, they were required to collect content live from their onsite locations on the day of the event. In India, reporting was live, nearly all done on scene. In Taiwan, all reporting was done earlier in the semester and posted to a microsite. This content was then translated by two UK Chinese-speaking students and posted the day of the event.

On the day of the project, many students reported live from the field. CSUN and ACJ students used phones or tablets and their personal social media accounts. NU students used these as well as compact video cameras and digital voice recorders, in which case they then returned to campus with the raw material to be edited in a temporary newsroom.

APPLICATIONS

Structure of the Project

Two main mechanisms were used to bring the content of this group of 127 students together: shared curation of a central Twitter account (@PopUpNewsroom) and a shared hashtag selected by the students (#LivePoverty), although several other hashtags (such as #childmedia and #elderlymedia, which were tied to Indian subgroups reporting on those topics) were also adopted. The hashtags allowed content to be followed and aggregated by students in different locations who did not actually know each other. They retweeted some of the content with the @PopUpNewsroom account, but due to limits set by Twitter of 1,000 tweets per day, not every tweet was reposted. Over 48 hours, #LivePoverty gained 386,541 impressions, meaning the tweets were delivered to that many Twitter accounts. These numbers were based on TweetReach analysis.

The main Twitter account @PopUpNewsroom was staffed by students from each university at designated times. Twitter coverage was also aggregated on the front pages of two websites (PopUpNewsroom.net and PopUpNewsUK.net) using RebelMouse, a social media aggregator. This tool automatically collects content from multiple forms of social media (Twitter, Facebook, YouTube, Pinterest, and so on), searching for hashtags the account holder sets. This served as an additional means of pulling together content from all participants, creating a global collection of student reporting.

The U.S. team launched the coverage in the late afternoon on Friday, November 15, Pacific Standard Time, staffing the main Twitter account @PopUpNewsroom and sharing content being created in Los Angeles and Chennai. After two hours, they passed control of the main Twitter account to India, where students produced their own coverage and retweeted their own and other schools' coverage. Five and a half hours after India came on line, Britain came in, producing their own content and retweeting U.S. and India material and posting the content from Taiwan. When India signed off, Britain took over until the United States came back on at around 5 p.m., Saturday, November 16, UK time (Greenwich Mean Time). CSUN students then retweeted links to the Indian, UK, and Taiwan content as well as producing their own.

Each university designed its own operation to produce coverage that matched each program's goals. For example, the U.S. team (15 students) concentrated on live field reporting through their own phones and social media accounts. Students rotated curation duties in a command center, which was a campus seminar room (not a student media space or lab) where students used either laptops, tablets, or phones to aggregate other students' content. None of their content was edited.

The Indian team (60 students) broke into teams based on subtopics such as LGBT, the elderly, and others. Those groups first aggregated student work via designated Twitter accounts; they also used other accounts to retweet content from each of these. Some of this work was done at the campus computer lab.

The UK team (52 students) established a steering committee of eight and divided into six teams, each taking a different topic: education, health, charities, politics, housing and homelessness, art and culture. But the boundaries between these dissolved and other issues were adopted as the project progressed. The UK students alternated between reporting from the field and taking editing duties in a command center, reposting tweets and editing and posting content prepared earlier. Some used university PCs and others used their own laptops. On the day, one or two students from India, the United Kingdom, and the United States communicated with their opposite numbers in the other countries, but time differences made this difficult, and contact mainly took place when one group handed over control of the main Twitter account or took it over from another.

The researchers conducted (a) a post-project survey of participating students combined with (b) an assessment of student interviews. An electronic survey was distributed to 127 participating students after the reporting project ended in November 2013. Of these, 71 students responded (not every question was answered by all 71). Participation was voluntary and anonymous and had no connection with students' grades. The survey was mainly intended to document the tools and applications the students used to report and how they responded to the project (which social media application the students used, which they used the most, which tool [such as phone, camera] they used and which they used the most). It also included demographic questions: the student's nationality, their previous reporting experience, and which university they were attending. The survey also posed an open-ended question: "What did you learn reporting for the Pop-Up Newsroom?"

The second data set consisted of 21 interviews that student participants in the United States and the United Kingdom conducted with other students from those countries and from India. Eight students from the Indian program, seven from the United Kingdom program, and six from the United States program were interviewed. Each student interviewed a student from a different country. All interviews with one exception were conducted through email; one interview was done via Facebook chat. Students were supplied with a common set of four questions and could add an additional question if they chose. Both the open-ended survey question and the interviews were analyzed qualitatively with thematic categories generated deductively from this data. This analysis was done through the constant comparative method in which the texts were read and reread, with each reading generating tentative categories in an ongoing process until the data were reduced to a set of key categories as reported below.

Findings

Experience. Most of the 71 students who participated in the survey were inexperienced as journalists. Nearly two thirds (65%) reported that their only previous journalism experience was through journalism classes, student news media, or internships, while another 20% lacked even student journalism experience prior to reporting for Pop-Up Newsroom. Eleven percent had one to three years of professional work, and 4% had worked in professional journalism for four or more years. This meant that most of the students would have experienced little to no direct newsroom socialization prior to their participation in Pop-Up Newsroom. These results cannot account for potential socialization via consumption of news media or popular culture representations of journalism practices.

Nationalities. While many students taking the survey were the citizens of the country where the four programs were based, a sizable minority came from 12 additional countries. The largest number of students was from India (25), followed by China (13), Taiwan (9), the United States (7), and the United Kingdom (4) (all were locations of the participating schools with the exception of China). Countries represented by students studying outside their home nations included Japan (2), Germany (2), and one student each from Spain, France, Bangladesh, Sri Lanka, Nepal, Denmark, Malaysia, Eritrea, and Kazakhstan. Thus, of those taking the survey, 37% were not from the country they were reporting from. While not accounted for in the assessment here, the international diversity of students in such a project would be important to examine in future studies.

Most Used Tools and Applications

RQ1: How do journalism students negotiate practicing journalism as a networked process rather than the traditional production of a linear, polished product? In regard to this question, data were gathered on the students' use of social and mobile equipment and devices.

The most used reporting device was the smartphone (74%); this is unsurprising as the Pop-Up Newsroom was based around the idea that students would report live in the field, collecting information and conducting interviews using mobile devices. Other tools students reported using most often included laptop (8%), desktop (7%), tablet (6%), digital camera (4%), and video camera (1%). There were variations between the schools. For example, all of the U.S. program participants indicated that a smartphone or tablet was their most used tool; in Taiwan, half of the students reported their most used tool was the digital camera. This likely reflected the emphasis on all mobile reporting at the U.S. school.

As they reported, many students tweeted their interviews, photos, and so on through their personal Twitter accounts. Thus, the finding that the most used application was Twitter was again to be expected, with more than two thirds (64%) reporting it as the tool they used the most. Other most used applications included Facebook (16%), Weebly (10%), YouTube (6%), Infogram (3%), and Instagram (1%). Again, there were variations by country, with Twitter the most used tool by 100% of the U.S. program participants, whereas it was most used by a little over half of the UK program participants (52%), many of whom favored uploading their finished texts to the project website, and Facebook (56%) was most used in Taiwan. The latter two programs incorporated an emphasis on a standing website to host their content; while the U.S. program also had such a site, it was not used to post standalone stories but rather to host the RebelMouse aggregations.

Students' Responses to the Project

Most students tended to describe their experiences in terms of how they experienced using new tools in their reporting and/or in terms of the ways the project made them aware of new ways to share their work. A final set of answers fell into learning the basics of reporting (such as how to interview); however, this analysis concentrates on the first two categories.

Becoming Mobile Journalists. Under this category, students emphasized the ways using their mobile phones and social media applications such as Twitter showed them the values of being multi-skilled and made them effective, faster reporters.

Speed. Repeatedly, students stressed the speed of mobile reporting, describing themselves learning "how to gather news and disseminate [it] almost in an instant," and engaging in "instant decision making, working at great speed." In other words, they believed they were performing what Van der Haak, Parks, and Castells (2012) describe as "journalism at internet speed" (p. 2933).

Multi-skilled. Some students noted that using mobile phones to report made them aware that they could carry out multiple roles—they could be converged journalists. As one said, she learned "to be an all-rounder journalist who can report, click pictures, capture videos, meet deadlines." Their comments also reflected that they realized the forms were merging: "I'm used to these roles being separate (reporter/photographer/videographer)." These were important comments because they reflected arguments by researchers such as Bardoel and Deuze (2001) that networked journalism relies on converged media to exist.

Interestingly, however, no student questioned the effectiveness of nonprofessional tools in the field; instead, like this one, they tended to note that "a smart phone is enough for efficient reporting and also to produce a quality end product." This was repeatedly cited as the most important thing they learned in the project: "We can report live with no big equipment in our hands like cameras and other stuff. But still can tell stories in a most effective way."

App Journalism. Beyond what is traditionally conceived as converged journalism practices, they further emphasized the importance of the applications they were using, with some reporting that the most important part of their experience was that "I learnt to use different mobile apps." Indeed, their answers suggested mastering apps was viewed as a crucial element of many students' experiences. One noted that "I didn't need to take out my pen and notepad even once. Instead, I was constantly on my phone, tweeting, whatsapping, instramming [*sic*]."

That is, some students began to create journalism directly from these social media tools, which arguably suggested that the tools might play an increasing role in socializing journalists, particularly beginners. For example, a student highlighted learning to meet the Twitter character limits: "How to . . . control the words (for Twitter limits the words I could say)." While brevity has long been a key journalism skill, this experience suggested students were learning a particular form of brevity dictated by Twitter. The Twittersphere and its constantly evolving unwritten practices and "rules" clearly socialized the tool's users. However, it was observed that some UK students, although they expressed appreciation of the utility of the smartphone as a reporting tool, actually used compact camcorders and digital voice recorders when reporting in the field and returned to campus-based facilities to edit the material, uploading it to a website, and tweeting a link to the content. This indicated a reluctance among some to rely on the social platform as a tool for primary reporting, and a tendency to prioritize the production of a polished text over immediacy and interactivity offered by social media.

Technology Challenges. While technology was seen as making their work possible, many students repeatedly suggested that technology was also their biggest challenge, in particular the tendency of the applications to quickly run down their phone's batteries: One noted that "I had to wait in a coffee shop for nearly an hour getting my phone charged." Some Indian students carried laptops with them to recharge their phones. India students also reported "connectivity problems when reporting from remote areas."

Dissemination and Collaboration

RQ2: How do journalism students negotiate the absence of spatial and symbolic newsroom boundaries, which characterize industrial models of journalism? In regard to this question and the students' reflection on their work, they were asked to consider their engagement and interaction with others. Beyond technology-related comments, students repeatedly emphasized the potential global dissemination of content and engagement with others. Both of these patterns suggest that the student journalists were not necessarily thinking of themselves simply as top-down deliverers of one-way information but rather saw how their work was embedded within networks of other reporters and audiences.

Widespread Dissemination. Students reported that social media enabled them to reach far beyond their own locality: "Global popupnewsroom [*sic*] fascinated me in terms of how quickly information could be transferred and communicated across to other countries via the Internet." Some students identified the ability of networked journalism to bring oftentimes-ignored voices to broader attention: "We can reach out to the world and show them poverty at the grass-root [*sic*] level."

They also believed that social media had the ability to gain the attention of those with the power to enact change.

Mobile Newsroom. Just as students identified expanded spatial distribution of their work, some said that they came to understand that news could be produced anywhere, that the "newsroom is not a specified area. Reports can be generated from the venue when and where the issue takes place." As one student noted, "I can cover any event live from anywhere," while another said that Pop-Up could "handle [any] matter, any time, any where [*sic*]."

Collaboration and Engagement. Students also described themselves as part of a network of reporters working "independently" to produce work with others. Some specifically took note of the ways social media brought them into greater contact with others, including potential audiences for their work. A student said she learned that it wasn't enough to simply send out content and hope it was seen: "In order to reach a public, you need to prepare the public. You need to start before the event itself and supply information, stories, promote, etc. If you don't do that, nobody is going to read what you report on the day." Others said, "I learned that we can engage/collaborate with the public, organizations and other citizen journalists in many ways." As one student summed up the experience: "The boundaries of journalism have changed to incorporate many new areas including social media, engaging communities and crowd sourcing. These used to be scary things for me to think about since I didn't fully understand them, but now I've learned these are great tools that will enhance the reporting process and aid the network of gathering and transmitting information."

DISCUSSION

A range of social, economic, and technological changes are combining to reshape the way journalism is practiced around the world. The spread of participatory culture and the digital tools that enable it can be viewed as threats to the traditional way of teaching journalism or they can be seen as unprecedented openings to re-imagine what the next generation of reporters is being taught. As evidenced in the creation of Pop-Up Newsroom, the rise of social media and the ability to produce media content under almost any circumstances can be seen as a paradigm-changing opportunity for educators. Embracing these changes, the Pop-Up project offered a new model for teaching journalism that encouraged students to spend more time in the field, reporting live from the communities they covered and even in collaboration with students working similar angles halfway around the world. This project is offered as an alternative to the traditional teaching space: a student newsroom, which is often isolated from the citizens whom journalists claim to represent and a place where students can find themselves socialized into professional routines that in our rapidly evolving field may lose their relevance before they can even graduate. The project was able to accommodate a diverse range of modes of engagement, so that students and instructors in different countries approached it in different ways. In Taiwan, students favored a more traditional mode of journalism practice relying on compact camcorders rather than smartphones. But their polished media texts were uploaded to Facebook for dissemination in the social sphere, and dissemination was reinforced by UK students adding translations from Mandarin to English and tweeting links to those texts, emphasizing global interconnectivities. In the United Kingdom, some students used smartphones and relied on Twitter as primary reporting tools, prioritizing processes of networked interactivity, while others clung to traditional signifiers of professionalism and used more traditional journalistic tools to generate content that was edited on campus, rather than in the field, and uploaded to a website as polished texts. In the United States and India, students engaged most wholeheartedly with social platforms and personal mobile devices such as smartphones to position their participation in the field, and in their communities. But all of the students embraced both the local and the global perspectives of journalism and situated their practice, both empirically and conceptually, in a collaborative, networked context.

The project demonstrated that, in regard to RQ1, journalism students negotiated practice in terms of both process and product, and, given the opportunity, sought to generate and interact with networks using digital, mobile, media platforms and devices. They also adhered to traditional journalism practice by producing texts to place before an audience, but they did so in part to provoke interactions and generate these through Twitter and other social media platforms.

In regard to RQ2, some students saw no need for the spatial, social, and symbolic support of the newsroom, while others sought it out as a professional space to which they returned from the field, a space in which they processed and polished their products and from which they disseminated those texts. But all the students reflexively developed understandings about evaluating their practice in terms of collaboration and interaction with students in other countries, and within wider communities and throughout local and global networks.

No educator can offer a prescription for the future of journalism, but this project accommodates the contention by Deuze (2008) and others that rapid change is now a permanent condition of modernity and the most effective education is one that prepares students to work with instability rather than trying to resist it.

REFERENCES

Anderson, C. W. (2013). *Rebuilding the news: Metropolitan journalism in the digital age*. Philadelphia, PA: Temple University Press.

Anderson, C.W., Bell, E., & Shirky, C. (2012) *Post-industrial journalism: Adapting to the present.* New York, NY: Tow Center for Digital Journalism.

Baines, D., & Li, T. (2013). *Students teaching in the community: Building employability, social sustainability and "journalism as process."* A paper presented at the World Journalism Education Conference, Mechelen, Belgium, June 3–5. Retrieved from http://wjec.be/downloads

Bardoel, J., & Deuze, M. (2001). Network journalism: Converging competences of media professionals and professionalism. *Australian Journalism Review, 23*(2), 91–103.

Beckett, C., & Mansell, R. (2008). Crossing boundaries: New media and networked journalism. *Communication, Culture & Critique, 1*(1), 92–104.

Berger, G. (2011). Empowering the youth as citizen journalists: A South African experience. *Journalism, 12*(6), 708–726.

Castells, M. (1996). *The rise of the network society: The information age: Economy, society, and culture.* Malden, MA: Blackwell.

Castells, M. (2007). Communication, power and counter-power in the network society. *International Journal of Communication, 1*(1), 238–266. Retrieved from http://ijoc.org/index.php/ijoc/article/view/46/35

Deuze, M. (2007). *Media work in a digital age.* London, UK: Polity Press.

Deuze, M. (2008). The changing context of news work: Liquid journalism and monitorial citizenship. *International Journal of Communication, 2*, 848–865. Retrieved from http://ijoc.org/index.php/ijoc/article/view/290/197

Domingo, D. (2008). Interactivity in the daily routines of online newsrooms: Dealing with an uncomfortable myth. *Journal of Computer-Mediated Communication, 13*(3), 680–704.

Flew, T., & Wilson, J. (2010). Journalism as social networking: The Australian youdecide project and the 2007 federal election. *Journalism, 11*(2), 131–147.

Gans, H. J. (1979). *Deciding what's news: A study of CBS Evening News, NBC Nightly News, Newsweek, and Time.* Evanston, IL: Northwestern University Press.

Heinrich, A. (2011). *Network journalism: Journalistic practice in interactive spheres.* New York, NY: Routledge.

Holton, A. E., Coddington, M., & Gil de Zúñiga, H. (2013). Whose news? Whose values? Citizen journalism and journalistic values through the lens of content creators and consumers. *Journalism Practice, 7*(6), 720–737.

Karlsson, M., & Strömbäck, J. (2010). Freezing the flow of online news: Exploring approaches to the study of the liquidity of online news. *Journalism Studies, 11*(1), 2–19.

Lewis, S. C. (2012). The tension between professional control and open participation: Journalism and its boundaries. *Information, Communication & Society, 15*(6), 836–866.

McNair, B. (2009). *News and journalism in the UK.* London, UK: Routledge.

Mensing, D. (2010). Rethinking [again] the future of journalism education. *Journalism Studies, 11*(4), 511–523.

Robinson, S. (2011). Journalism as process: The organizational implications of participatory online news. *Journalism & Communication Monographs, 13*(3), 137–210.

Robinson, S. (2013). Teaching "journalism as process": A proposed paradigm for J-School curricula in the Digital Age. *Teaching Journalism and Mass Communication, 3*(1), 1–12. Retrieved from http://www.aejmc.net/spig/journal

Stencel, M., Adair, B., & Kamalakanthan, P. (2014). The goat must be fed; why digital tools are missing in most newsrooms. *Duke Reporters Lab.* Retrieved from http://www.goatmustbefed.com/

Thurman, N., & Myllylahti, M. (2009). Taking the paper out of news: A case study of *Taloussanomat*, Europe's first online-only newspaper. *Journalism Studies, 10*(5), 691–708.
Tuchman, G. (1978). *Making news; a study of the construction of reality*. New York, NY: Free Press.
Usher, N. (2014). *Making news at the* New York Times. Lansing, MI: University of Michigan Press.
Van der Haak, B., Parks, M., & Castells, M. (2012). The future of journalism: Networked journalism. *International Journal of Communication, 6*(16), 2923–2938. Retrieved from http://ijoc.org/index.php/ijoc/article/viewFile/1750/832
Van Dijk, J. (2012). *The network society*. Thousand Oaks, CA: Sage.
Wall, M. (2015). Change the space, change the practice? Re-imagining journalism education with the Pop-Up Newsroom. *Journalism Practice, 9*(2), 123–137.

ASSIGNMENT 9.1. GLOBAL POP-UP NEWSROOM POVERTY PROJECT

Preparation

1. Pick a topic. Log it into our shared Google Spreadsheet. Make sure your topics do not overlap.
2. Identify potential sources/stories. While you are allowed to do a pre-interview, you must be able to be on site the day of the Pop-Up, so you can report live.
3. Micro Literature Review. Locate two articles or reports related to your topic and summarize these in two pages. How will this inform your reporting?
4. Engagement. Share what you are doing through your own social networking sites. Engage with strangers who are interested in your topic by liking, following, @ them on Twitter and/or Facebook.
5. Evaluate a social media tool.
   a. Visit the tool's site. If they have a tutorial, watch/read it.
   b. Sign up for the tool and use it.
   c. Evaluate the benefits/obstacles of its use for the global project. Present your assessment to the class.

Day of Pop-Up

6. Reporting for Pop-Up Newsroom
   Pick one of two roles: Reporter or staff the Command Center, which monitors the content being produced, sharing and reposting it. These roles may swap out during the event.

Reporters:

- Work in reporting pairs or alone. If pairs, you must still produce your own Tweets.
- Use your cellphones (If you don't have a smartphone, check out an iPad from the Journalism Department.)

Original content	How frequently?	What should do/contain
Text (Tweet)	Minimum of 4 original Tweets + 3 retweets of classmates, 2 of strangers.	Use our hashtag, no typos, be accurate.
Photos	Minimum 2 (should be of different people).	Identify person by full name if relevant; use hashtag, no typos, be accurate.
Video or Audio	Minimum 1.	Identify person by full name if relevant; use hashtag, no typos, be accurate.

Command Center

- Learn how to use Twitter, RebelMouse, and Storify ahead of time. As you select materials to share, think about what would most engage people. Try to be representative.

Tool	How frequently?	
Twitter (retweets; no more than 10 every 30 minutes)	Tweets: 8–10 every 30 minutes. Use Favorites. Post to Facebook once every 30 minutes.	When RTing our reporters, correct their grammar/accuracy as necessary.
Use Storify or similar tool to aggregate content.	1 Storify or other tool per 90 minutes.	

Post Event

7. Reflection. Write a reflection on your experience at the class meeting after the Pop-Up is held. Questions to be supplied.
8. Cross-national interviews.

Conduct a brief interview via email, Skype, Facebook chat, Google Hangout, etc., with a participant from another country. (Names and questions will be supplied.) What do you learn from them about their experiences with the project?

CHAPTER TEN

Storytelling 2.0

Using Social Media Tools to Craft Multimedia Stories

LEIGH LANDINI WRIGHT

Journalism instructors of the past handed out story notes compiled from their own journalistic endeavors or from the pages of a workbook to teach students how to write a news story. The exercises became predictable and did not truly simulate the professional world experience of writing and reporting. With the advent of social media in the past decade, though, instructors had new and accessible tools available to enhance the traditional news writing exercises. Students were already engaged with tweeting, as evidenced by the Pew Internet Research Project's Social Media Update (Duggan & Smith, 2013) that found 31% of people ages 18–29 used Twitter and accounted for the age group with the highest usage. Moreover, 78% of Twitter's monthly users made their tweets via mobile devices (Bennett, 2014).

Students utilized a variety of social media tools such as Twitter, Vine, or Instagram to communicate with their peers. Instructors saw students tapping furiously into their phones or tablets during lectures and wondered what was so important that the students must type on small keypads instead of taking notes. Rather than banning mobile devices and computers as well as social media in the classroom, instructors used these tools to enhance instruction and provided examples of how professional media organizations used social media to report stories, engage with their audience, and create new methods of digital and social media storytelling.

Educators have experimented with social media tools to enhance traditional lecture-based instruction in recent years. It has become common to see students working on a class assignment on Twitter with a specific class hashtag or creating Vine videos ranging from public service announcements to promotional pieces.

They also used Storify as a multimedia storytelling tool or used Instagram as a video and photo storytelling service to supplement multimedia coverage. However, the usage of these tools required practice on the part of both the student and the educator in order to achieve a rich learning experience that allowed students to explore inquiry-based and active learning practices. Social media had to be taught as a hands-on tool with exercises that forced the student to explore and develop their storytelling voices and writing structures. Students benefitted from experiences that brought together the best exercises from a workbook or textbook with the digital tools they need to succeed in modern newsrooms.

Rooted within the theories of Web 2.0 (O'Reilly, 2007) and connectivism (Siemens, 2004), this chapter examines the usage of a variety of social media tools (Twitter, Vine, Instagram, and Storify) to enhance the multimedia storytelling experiences for journalism students and offers several examples, exercises, and student reactions from a journalism program of a southeastern regional university. The chapter also provides a series of best practices and applicable tips for instructors to consider using in the classroom to train a future generation of journalists who will use current as well as emerging social media platforms for storytelling in newsrooms that have increasingly become a digital-first structure with a heavy emphasis on using social media tools for audience engagement. The chapter is also designed to help journalists and media instructors, particularly those who teach writing and reporting skills, to build additional social media storytelling skills into their courses.

LITERATURE REVIEW

Social Media in the Newsroom

Today's journalists turn to their smartphones and tweet an update from a local government meeting, post a Facebook update, or post a photo or video to Instagram or Vine. Nearly 40% of U.S. journalists surveyed in 2013 reported they believed social media were very important to their daily tasks, and 53.8% said they used Twitter or other microblogging platforms to gather and report information (Willnat & Weaver, 2014). Additionally, The Oriella PR Network (2013) found 59% of journalists worldwide used Twitter, with higher usage rates found in the United States, the United Kingdom, France, Spain, Canada, and Australia.

Established in 2006, Twitter allows users to post 140-character messages, tweets, to their followers. Perhaps one of the best examples was a National Public Radio correspondent who used Twitter to report the story of the 2011 Arab Spring uprising. The correspondent collected information from the streets and tweeted up to a thousand times a day (Shephard, 2013). The correspondent explained that multiple tweets were used to instill context to the readers and to dispel rumors

(Shephard, 2013). The usage of Twitter as a news tool has continued to evolve. When Hurricane Sandy hit the East Coast in 2012, 34% of the tweets produced during the storm involved news organizations, government sources, and the public reporting on news and human interest (Guskin & Hitlin, 2012). News organizations used Twitter to report in real-time after the Sandy Hook, Connecticut, shooting in 2012 and the Boston Marathon bombing in 2013; however, in both cases, the media, as well as the general public, made mistakes in the haste to disseminate information (Beaujon, 2013; Reinwald, 2014).

As new social media networks began, reporters and photographers experimented with these for enhancing their storytelling and reporting tools. Journalists, particularly photojournalists, used Instagram as a way to communicate with the visual social media audience. For instance, the Associated Press photographers utilized the hashtag #APonthetrail to tag their Instagram posts, and CNN engaged with its audience during the 2013 inauguration to encourage viewers to upload their snapshots from the event (Laylin, 2014). The Guardian and BBC, for instance, launched Instagram projects in early 2014, ranging from a headline-type video package to behind-the-scenes packages captured in 15 seconds (Bartlett, 2014).

Vine, the six-second video looping service, launched in January 2013, and Twitter assumed ownership (Vine, 2014). Remarking on its first anniversary, Vine's founders wrote on their blog, "The creative community that has embraced Vine's short looping videos has shown that you can tell a whole story, make people laugh and even leave people speechless—in six seconds or less" (Vine, 2014, para. 1). News organizations use Vine as a tool to cover breaking news, to report events, to visualize data, and to show the behind-the-scenes footage (Edge, 2014). For instance, one Midwestern news station reporter said her newsroom used Vine to show the personalities behind the news as well as to find stories, using local hashtags (Beaujon, 2013). To respond to Vine's popularity, Twitter released a best practices guide for journalists on how to use the six-second service. Suggestions included showing breaking news, making promotional materials, stringing Vine videos together for a linear story, and creating a datavine to help break down numbers (Luckie, 2012).

As social media tools grew, former Associated Press correspondent Burt Herman spent a year as a Knight Fellow at Stanford University in 2009 where he devised the idea for Storify and decided to launch a curation service to help users make sense of the noise of social media (Fincham, 2011). The tool, dubbed Storify, began in September 2010, but Herman had to invite users until April 2011 (Fincham, 2011). By the end of 2012, Storify recorded 46,000 publishers producing 554,000 stories that garnered 367 million views (Wilhem, 2012).

With roughly half of all social media users getting news from platforms such as Facebook and Twitter and to a lesser extent Instagram at 13% and Vine at 9% (Holcomb, Gottfried, & Mitchell, 2013), it was not surprising that the number of newsrooms designating a staffer as a social media editor grew from 100 in 2012 to 184 in 2013 (Sasseen, Olmstead, & Mitchell, 2013). Additionally, the growth area for jobs in journalism appear in the digital realm. The Pew Research Center reported 468 digital outlets have produced 5,000 jobs (Jurkowitz, 2014). These digital outlets work with new storytelling techniques, particularly with multimedia and social media, aimed at the digital native generation (Jurkowitz, 2014).

Theoretical Framework: Web 2.0 and Connectivism

The two theoretical underpinnings of connectivism and Web 2.0 are combined to illustrate the tools involved with learning connections.

Web 2.0. Tim O' Reilly coined the theoretical concept of Web 2.0 in 2005. O'Reilly noted the participation factor of Web 2.0 through the use of tags (O'Reilly, 2007). These technologies encompass tools such as social networks (Facebook), media sharing platforms (YouTube and Flickr), and creative content such as blogs and microblogs (Twitter). These tools promoted and improved educational processes (Cohen & Duchan, 2012).

Web 2.0 has become both a platform on which innovative technologies have been built and a space where users shared content (Cohen & Duchan, 2012). Web 2.0 tools have been highly developed, and their role in education grew (Yakin & Tinmaz, 2013). This technology has facilitated online learning by increasing interactivity, active participation, and feedback mechanisms (Harrison & Thomas, 2009). However, the usage of Web. 2.0 in the classroom had challenges as well as advantages. Students became active learners who could access information and knowledge anytime (Harris & Rea, 2009), but learning required the usage of computers and social network spaces on which the information can be published and accessed by the general public (Harris & Rea, 2009). Thus, the usage of social media tools such as Twitter, Vine, Instagram, and Storify, among others, in a classroom was considered as a Web 2.0 tool since the posts can be accessed by the general public.

Connectivism. As the digital era marched into the classroom, Siemens developed the learning theory of connectivism in which learners connect to a network, share information, and connect to the network in a dynamic way (Siemens, 2004). Siemens suggested in the past 20 years that technology has restructured how people communicate, learn, and live (O'Bannon & Britt, 2012). In the classroom, educators observed this theory when students engaged with technology for assignments, shared the information, and established relevance to the content.

Within the connectivist model, learning happened when learners connected to a network to share and find information. The information found then helped the learners to change their thoughts based on the connections to the piece of information. The learners then shared their new information (Siemens, 2008). This type of learning connected with the usage of social media tools in which the students found information and then connected it to their assignments and writings.

The theory had roots in the premise of distributed knowledge. In the connectivist model, knowledge was distributed across an information network and stored in a variety of digital formats (Kop, 2008). Learning took place when cognition and emotions combined in the process, which then became cyclical. Learners connected to a network to share and find new information, modified their thoughts based on the new information, and then connected to a network to share these new discoveries and find new material. Learning not only was consumed but created (Siemens, 2008). Siemens had explained that understanding occurred when a learner became engaged in creating and re-creating his own learning network.

METHOD

An active-learning instructional model (Bonwell & Eison, 1991) was used for teaching writing and reporting skills at a southeastern regional university. In the active-learning instructional model, the students played an active role in their learning process through hands-on learning.

With an active learning teaching method, the instructor introduced the concepts of social media storytelling into a 200-level advanced newswriting class, but the real learning came when the students began working with the social media tools. The class examined writing and beginning reporting practices for a convergence-based journalism model, and thus, the students learned how to write and produce news for print, broadcast, online, social, and mobile media audiences. To achieve this, several assignments carried multiple media formats such as social media, multimedia, and mobile media as students learned how to report and write in a multiplatform environment. These types of assignments fell in line with the framework of Web 2.0 technologies and connectivism as students had to use social media tools and created their own network of learning.

In this particular model of learning, students were exposed to the concept through a lecture that used presentation-based software. However, social media skills were then demonstrated on both the classroom computer as well as a personal mobile device, and then students were allowed to try some test tweets. Once the students became familiar with Twitter for news purposes, they watched a brief video lecture given by a celebrity at a college commencement ceremony. They then mock live-tweeted the event while the instructor displayed the tweets on the overhead projector and offered real-time feedback and instruction. This in-class exercise demonstrated the ease of using Twitter combined with the challenges of live-tweeting an actual speaker.

Students then were instructed in the art and practice of social media curation through a demonstration of how to use Storify. Tweets from their mock live tweeting exercise were pulled together when the instructor found the class hashtag posted on Twitter. Those posts were then curated into the Storify. Narrative writing techniques were shown to the students so that they could tie their tweets together. Once the instructor demonstrated the skills, the students tried the new curation techniques of Storify themselves by assembling the mock live-tweeting exercise.

With only 10 to 15 students in a class any given semester, the instructor spent additional time to perfect their tweets and Storify techniques before sending them to an assignment. The class at the 200-level, primarily sophomores and a few juniors, examined convergent media writing, while the 300-level or intermediate reporting class of 10 to 15 students, primarily juniors, examined how to write and report for print media.

Using social media tools such as Twitter in a journalism classroom required practice as a pedagogical tool. The tool needed to go beyond the surface and look into media richness, social presence and identity, and social information and processing (Hosterman, 2012). In a large lecture class, such as an Introduction to Mass Communication, an instructor used Twitter for his students to interact with a guest speaker, discuss class ideas in real-time, and conduct small-group exercises within the larger group (Tyma, 2011). However, in smaller classes, such as writing classes, instructors used Twitter to tweet content before class or used it as a tool to teach about current events (Fox, 2013). When used effectively, Twitter achieved a content relevance and enhanced student engagement (DiVerniero & Hosek, 2013).

With all the tweets and posts, one might wonder how to make sense of the tools in the classroom. Curation became the answer to making sense of the noise of social media. Curation referred to the practice of people gathering the social media elements and assembling them via software or a website into a readable format (Rosenbaum, 2010). Examples of this type of service include, but are not limited to, Storify, Scoop.It, and RebelMouse.

The instructor of these courses chose to use Storify for its ease of operation and ease of seamlessly integrating social media accounts. When Storify was used as a digital narrative, the writer crafted a brief headline or title for the post and then wrote an introductory paragraph, such as a lead. The drag-and-drop functions that connect apps (such as Twitter, Facebook, Instagram, web links, GIFs, YouTube, Flickr, Google Plus, or SoundCloud) allowed the writer to incorporate a variety of visual, audio, and social media links and then provide context within the narrative spaces of the text boxes.

Curation using Storify within the classroom provided a student-driven, creation-driven, collective exploration into integrated multimedia storytelling (Cohen & Mihailidis, 2012). Educators also used this tool as a way to engage students in discussions about media framing as well as ethics as they curated and produced their stories (Cohen & Mihailidis, 2012).

APPLICATIONS

Based on the current literature and best practices of using social media in the classroom, the instructor devised several social media storytelling exercises to reinforce the concept of active learning. Prior to the implementation of these exercises, the instructor discussed the importance of the news media using tools such as Twitter, Storify, Vine, and Instagram to connect with audiences and distribute their messages over a connected network. Students examined the social media accounts of selected national and regional television stations and newspapers and determined how these outlets used social media. They then discussed how effective the social media accounts were as storytelling tools.

Students also discussed the merits of live tweeting at meetings and lectures and contrasted the merits with the challenges. Students mentioned that trying to take notes with a pen and notebook became more challenging if they had to tweet at the same time. Several students said the tweeting could become the basis for their own notes from an event.

Live Tweeting and Storify

In the spring semester of 2013, seven students in a 200-level journalism class attended a university-wide lecture and live-tweeted the event with 15 tweets minimum, but they could post more if they wished (see Assignment 10.1). They were required to use the official event hashtag, and all tweets had to conform to Associated Press style and proper English usage. Students could not use abbreviations or slang. The tweets had to relate to the event. The Twitter stream was monitored throughout the evening and the instructor engaged in tweeting the lecture as a way to reinforce the model of effective concise tweeting.

The live-tweeting exercise helped the students, many of whom had not yet covered a live speech, develop concise note-taking skills. The students quoted the speaker, and since Twitter allowed only 140 characters, the quotes did not become bogged down with extraneous information. The students also wrote descriptions about the crowd interaction and atmosphere. As students tweeted, they noticed people at the event that followed the official hashtag retweeted their tweets or marked them as favorites.

After the lecture, the students compiled their tweets and photos into a Storify due at the next class session. Students were allowed to write their Storify in any story structure and most chose to write the piece as a straight-news inverted pyramid style. Their narrative spaces included remarks about the crowd size, the questions the crowd asked, and background or context about the speaker. By making the assignment a live and interactive deadline experience, students gained the knowledge of working on a deadline.

Two students experienced technology issues related to either their smartphone data plan limit or an overloaded Internet network and completed their tweets immediately after the event when they returned to their rooms. These types of issues happen in professional situations and students must be prepared for technology problems and adapt accordingly.

Incorporating Additional Social Media Platforms with Storify

After two years of combining Twitter and Storify for live-tweeting assignments and other social media activities, the instructor decided to try something different that incorporated more social media platforms. Students in the fall 2014 advanced newswriting class received an assignment to report a news story via social media with posts in Twitter, Vine, and Instagram and then write the story in Storify (see the TwiVineAGram, Assignment 10.2). Students had the option to choose their narrative story structure based on their subject and their interviews.

Students selected from three topics of news coverage: Black Friday shopping madness, advice from upperclassmen to underclassmen about surviving the week before final examinations that some colleges refer to as "Dead Week," and a guide to understanding the school's basketball obsession. Students were given a list of suggested questions. However, students developed their own questions for their sources, which allowed them to test the theory of active-based learning, along with using social media to draw connections to their learning and comprehension patterns. Although guidelines were given about the types of social media to use, students decided the best method in which to tell the story.

Since the class learning outcomes touched on the convergent nature of today's media environment, the instructor required students to think beyond words. They had to use a combination of photos and videos to tell the story visually and had to convey the information to the social media audience via Twitter. Based on the emerging popularity of Instagram and Vine, the instructor asked the students to capture the images on their smartphones or tablets.

The assignment had to combine the interviewee's answer or quote as well as the reporter's observation with a visual. The instructor set up a class hashtag to follow the students' progression and to help the class find relevant tweets for their own Storify stories. The instructor set up parameters for the students to write the story such as each story had to have a clever headline of four to eight words to achieve Search Engine Optimization, an attention-grabbing lead, transitions, and background or context, as well as the tweets, Vine, Instagram, and photos placed as quotations or individual anecdotal storytelling elements.

After students completed the assignment, they viewed the Storify posts from their peers. Several students then used their new Storify skills for completion of a multimedia group project at the end of the semester. Student feedback indicated

Storify could be considered a valuable social media storytelling device that could be embedded into a website or left as a stand-alone web link.

The instructor devised a 20-point rubric divided into five categories (headline, lead, content, focus, and mechanics) with four points as the maximum in each category. The content required the students to use the tweets and visuals along with crisp narrative to tell a complete story for the maximum points in that category.

After two years of using Twitter and Storify as tools to enhance storytelling, evaluation of the technique showed results as 90% positive. Ten randomly selected students who took the class participated in email interviews about their perceptions about the usage of Twitter and Storify as tools to enhance their writing and storytelling skills. Only one student indicated that the combined tools were a "waste of time," and the same student found the entire process "frustrating." Other students indicated that the ease of use of the tools helped them to think about multimedia stories. Comments included, "It helped me grasp the concept of multimedia storytelling tools. In my internship, I had to adapt and learn new storytelling tools similar to Storify and Twitter. I felt much more confident of my ability to navigate new sites." "These experiences taught me to be more efficient in using social media tools. The hands-on style of this activity made the lessons a much more memorable experience. During my internship, my experiences with these activities allowed me to be conscious enough to provide innovative ideas to improve my workplace environment."

Although Storify had a positive reaction from students, the usage of Twitter did not. Five of the 10 students said Twitter had not helped them make their writing more concise, and one student indicated that Twitter seemed too choppy and that it fragmented news sources. The five students who agreed that Twitter had helped their writing become more concise said it had forced them to eliminate words.

Additional Tools: Storify for Additional Readings

Because Storify allowed for web links as well as social media posts, the service could be used to pull together timely articles culled from both news sources and journalism organizations such as poynter.org, The Nieman Lab, and the Shorenstein Center. Tweets as well as web links were curated to accompany lessons about government, education, legal issues, and ethics in a 300-level reporting class. The analytics of Storify showed the number of times the post was accessed, and through anecdotal evidence, the instructor determined that most of the students read through the post at least once prior to class. Additionally, if the instructor embedded the Storify readings link into a learning management system (LMS), several such systems include an analytics portal for faculty to review student progress. Although students could read the technical aspects of coverage of these specific beats through textbook readings, the additional social media readings provided current and relevant examples for in-class analysis and discussion. Readings could have been pulled together as multiple links in the campus learning management system (CLMS), but the Storify allowed for one link within the LMS, and the Storify could be preserved and modified for future classes. Additional comments could be typed into the text boxes for issues that continue from one semester to another, and these comments helped to guide the students' comprehension and retention of these articles for class discussion. Updated academic or trade publication website articles also could be added as issues arise that demand attention in an academic setting.

Best Practices for Storify

Storify offered writers an opportunity to do more than simply drag and drop tweets and visuals into an embeddable format suitable for placement into a blog or a news website. The narrative spaces of Storify or the boxes between the tweets allowed writers to offer rich interpretation, context, background, or transitions, much like writers achieve in traditional news writing. Writers thought of the tweets as quotations, but the narrative spaces provided context and background to enhance the audience's understanding of the subject.

Educators might consider the list below as tools for their best practices of using Storify in the classroom as an instructional method for teaching social media writing and reporting skills.

1. Provide students with examples from professional news organizations such as NBC News, the Moscow Times, the Tampa Bay Times, NPR News, and the Wall Street Journal. By showing students the professional usage of Storify, the skeptical students will see that social media curation serves a legitimate journalistic purpose.
2. Require students to analyze Storify curations from various news organizations. This will force them to read regularly and consume news as well as notice how a news organization structures the social media curation post.
3. Recognize that although these students classify as digital natives (Prensky, 2001), some students did not regularly engage in social media and had an aversion to it. Consider asking students to create specific social media accounts for their journalism classes. By creating a separate account for their academic life, the students can show a potential employer any of their successful social media class assignments.
4. Model appropriate Storify techniques by writing and posting class materials that involve social media and curation services. For instance, consider posting additional course readings as web links embedded within the Storify and provide additional context in the narrative spaces.

5. Create a rubric that touches on the key points that needed to be emphasized. For a class that involved visual techniques, the rubric might include criteria about the number of photos or videos, either through Vine or Instagram, that must be included. For a writing class, the rubric might include a specific number of tweets that must be used along with a description about the type of narrative expected.
6. Tell students not to put placeholder words like "headline goes here" in the headline box as Storify will save the post with the URL extension with those words. The students need to write the real headline first.
7. Provide opportunities for students to practice and improve their skills with Storify. The tool can be introduced in beginning newswriting or reporting class as a way to capture tweets, and then the students can build upon the curation skill as they add context as social media stories in advanced newswriting and intermediate reporting classes. Students in the capstone (or senior-level) class could use Storify as one of the tools in a comprehensive multimedia package to tell a public affairs story and then embed the Storify into the course website.
8. Although social media allows for a more casual writing tone, students should still be held to the high standards for writing mechanics and structures as they would for a traditional print or broadcast story. The tweets and Storify posts must include proper Associated Press style and mechanics. Points should be deducted for incorrect grammar or unsuitable abbreviations.
9. Students must verify tweets and their information before posting into Storify, just as they would for a traditional news story. Students also need to consider any ethical implications just as they would for a print or broadcast story. Two useful tools for these endeavors include *The Verification Handbook* and the *NPR Verification Handbook.*
10. Instructors and students must keep in mind that not every story is suitable for social media writing. Some stories, such as routine write-ups about wrecks, do not lend themselves to social media storytelling. Stories that work well for Storify include coverage of special events, meetings, court cases, athletic events, major breaking news, and multifaceted feature stories. The Storify must have a strong social media element (a variety of informative and interesting tweets, good visuals, strong web links for background) in order to achieve a good curation.

DISCUSSION

Using social media tools such as Twitter and Storify in the classroom helped students develop an awareness of storytelling skills that stretched beyond print or broadcast platforms. News organizations have worked for several years to move toward more digital and social media storytelling tools and skills.

For some students, particularly those not comfortable with video tools, the accessibility and ease of use of smartphones or tablets made this transition to multimedia storytelling more accessible. The fact that they picked up their smartphone device and shot a series of visuals helped them to build the confidence they needed to embrace multimedia storytelling without using complicated video equipment. The software and the social media tools could evolve in the future, but the usage of the current set of tools should help these students become comfortable with telling stories in a social media and online news space.

Additionally, these social media tools such as Instagram and Vine have helped students to become more comfortable with shooting and producing video on mobile devices. These mobile devices have been used by industry professionals in the following ways: recording video, recording and filing audio, capturing images discreetly, producing live reports from the field, and filing copy (Belmaker, 2013); the trend toward more mobile and social journalism should continue well into the future.

The fact that students applied these lessons in their internships offered validation that news organizations took social media and curation tools seriously. Even if the brand of the tool changes, the digital skill set should help these students adapt to other tools and platforms.

However, even though instructors assumed that the digital generation has grown up knowing how to use social media, it was not an accurate perception as students deemed as digital natives still required instruction about social and multimedia platforms (Bor, 2014). These skills needed to be scaffolded, or shown and taught in a progression, into the classes to show reluctant digital natives why they needed to use social media in the classroom (McLoughlin & Lee, 2010). Based on anecdotal evidence in the researcher's classrooms, some students arrived in college classrooms without knowledge of any social media because their parents didn't allow the usage, and a few students had no interest in ever using social media, even after repeated lessons and assignments. The scaffolding that McLoughlin and Lee (2010) suggested could have helped overcome the challenges of students not wishing to engage with social media.

However, with digital media showing the best signs of growth in the media industry with an estimated 5,000 jobs at 468 outlets (Jurkowitz, 2014), instructors who wanted their students to find jobs after graduation would be well-served to engage students with effective active-learning teaching models that use social media as a storytelling method as well as teaching the traditional methods of news gathering and reporting. Regardless of if the lessons involved emerging reporting techniques or traditional reporting techniques, instructors must use effective methods of practicing these new and emerging skills since the practice of journalism required learning by doing. Exercises such as the ones outlined in this chapter helped students practice and perfect their social media storytelling techniques. Furthermore, instructors should continue to stay engaged with new and emerging social media platforms to prepare students for the jobs of the future.

REFERENCES

Bartlett, R. (2014, February 4). How the BBC and Guardian are innovating on Instagram. Retrieved from http://www.journalism.co.uk/news/-how-bbc-guardian-innovating-instagram/s2/a555771/

Beaujon, A. (2013, May 8). How journalists can get serious content from Vine. Retrieved from http://www.poynter.org/latest-news/mediawire/212892/how-journalists-can-get-serious-content-from-vine/

Belmaker, G. (2013, January 18). 5 ways journalists can use smartphones for reporting. Retrieved from http://www.poynter.org/news/media-innovation/200418/5-ways-journalists-can-use-smartphone-apps-for-reporting/

Bennett, S. (2014, June 9). Facebook, Twitter, Instagram, Pinterest, Vine, Snapchat: Social media stats 2014 [infographic]. Retrieved from http://www.adweek.com/socialtimes/social-media-statistics-2014/499230

Bonwell, C. C., & Eison, J. A. (1991). Active learning: Creating excitement in the classroom. 1991 ASHE-ERIC Higher Education Reports. *ERIC Clearinghouse on Higher Education*, 20036–1183.

Bor, S. E. (2014). Teaching social media journalism: Challenges and opportunities for future curriculum design. *Journalism & Mass Communication Educator*, *69*(243), 255. doi:10.1177/1077695814531767

Cohen, A., & Duchan, G. (2012). The usage characteristics of Twitter in the learning process. *Interdisciplinary Journal of E-Learning and Learning Objects*, *8*, 149–163. Retrieved from http://www.informingscience.org/Journals/IJELL/Overview

Cohen, J., & Mihailidis, P. (2012). Storify and news curation: Teaching and learning about digital storytelling. Retrieved from http://socialmediatechnologyconference.com/wp-content/uploads/2013/04/SMTCProceedingsFinal2012-1.pdf#page=27

DiVerniero, R., & Hosek, A. (2013). Twitter as a classroom tool: Exploring the use, benefits, and downfalls from the perspectives of instructors and students. *The Journal of Social Media in Society*, *2*(2), 48–75.

Duggan, M., & Smith, A. (2013, December 30). Demographics of key social networking platforms. *Pew Research Centers Internet American Life Project RSS*. Retrieved from http://www.pewinternet.org/2013/12/30/demographics-of-key-social-networking-platforms/

Edge, A. (2014, January 27). 5 ways newsrooms used Vine in its first year. Retrieved from http://www.journalism.co.uk/

Fincham, K. (2011). Review: Storify. *The Journal of Media Literacy Education*, *3*(1), 56–60.

Fox, K. (2013, November 12). Twitter in the classroom. *New Chalk Talk*. Retrieved from http://www.aucegypt.edu/llt/clt/ChalkTalk/Documents/Volume%2013/Chalktalk%20Vol13_Issue%201.pdf

Guskin, E., & Hitlin, P. (2012, November 6). Hurricane Sandy and Twitter. Retrieved from http://www.journalism.org/2012/11/06/hurricane-sandy-and-twitter/

Harris, A. L., & Rea, A. (2009). Web 2.0 and virtual world technologies: A growing impact on IS education. *Journal of Information Systems Education, 20*, 137–144.

Harrison, R., & Thomas, M. (2009). Identity in online communities: Social networking sites and language learning. *International Journal of Emerging Technologies & Society*, *7*(2), 109–124.

Holcomb, J., Gottfried, J., & Mitchell, A. (2013, November 14.). News use across social media platforms. Retrieved from http://www.journalism.org/2013/11/14/news-use-across-social-media-platforms/

Hosterman, A. R. (2012). Tweeting 101: Twitter and the college classroom. In H. Noor Al-Deen & J. Hendricks (Eds.), *Social media: Usage and impact* (pp. 93–110). Lanham, MD: Lexington Books.

Jurkowitz, M. (2014, March 25). The growth in digital reporting. Retrieved from http://www.journalism.org/2014/03/26/the-growth-in-digital-reporting/

Kop, R. (2008). *Web 2.0 technologies: Disruptive or liberating for adult education?* Adult Education Research Conference. St. Louis, MO. Retrieved from http://www.adulterc.org/Proceedings/2008/Proceedings/Kop.pdf

Laylin, C. (2014, February 27). How Instagram could change journalism. Retrieved from http://newszou.com/how-instagram-could-change-journalism/

Luckie, M. (2012, September 20). Best practices for journalists. Retrieved from https://blog.twitter.com/2012/best-practices-for-journalists

McLoughlin, C., & Lee, M. J. (2010). Personalised and self regulated learning in the Web 2.0 era: International exemplars of innovative pedagogy using social software. *Australasian Journal of Educational Technology*, *26*(1).

O'Bannon, B. W., & Britt, V. G. (2012). Creating/developing/using a wiki study guide: Effects on student achievement. *Journal of Research on Technology in Education*, *44*, 293–312. Retrieved from http://iste.org/jrte

O'Reilly, T. (2007). What is Web 2.0: Design patterns and business models for the next generation of software. *Communications & Strategies* (1), 17.

Oriella PR Network. (2013). The new normal for news: Have global media changed forever? Retrieved from http://www.oriellaprnetwork.com/sites/default/files/research/Brands2Life_ODJS_v4.pdf

Prensky, M. (2001). Digital natives, digital immigrants. *On the Horizon*, *9*(5), 1–6. doi:10.1108/10748120110424816

Reinwald, C. (2014, April 18). What Twitter got wrong during the week following last year's Boston Marathon. Retrieved from http://www.boston.com/news/local/massachusetts/2014/04/18/what-twitter-got-wrong-during-the-boston-marathon-bombing-week/ZOYLJpEydYgJ8UYNUT674H/story.html

Rosenbaum, S. (2010, May 3). Why content curation is here to stay. Retrieved from http://mashable.com/2010/05/03/content-curation-creation/

Sasseen, J., Olmstead, K., & Mitchell, A. (2013). Digital: As mobile grows rapidly, the pressures on news intensify. Retrieved from http://stateofthemedia.org/2013/digital-as-mobile-grows-rapidly-the-pressures-on-news-intensify/

Shephard, M. (2013, February 23). Q&A: NPR's Andy Carvin on tweeting and tweeting and tweeting. *The Toronto Star*. Retrieved from http://thestar.com

Siemens, G. (2004). Connectivism. A learning theory for the digital age. Retrieved from http://www.elearnspace.org/Articles/connectivism.htm.

Siemens, G. (2008). Description of connectivism, connectivism: A learning theory for today's learner. Retrieved from http://www.connectivism.ca/about.html

Tyma, A. (2011). Connecting with what is out there!: Using Twitter in the large lecture. *Communication Teacher, 25*, 175–181. doi:10.1080/17404622.2011.579911

Vine. (2014, January 23). Our first birthday: A year on Vine. Retrieved from http://blog.vine.co/post/74314496911/our-first-birthday-a-year-on-vine

Wilhem, A. (2012). Storify users posted 554,000 stories in 2012 which were viewed 367 million times. Retrieved from http://thenextweb.com/insider/2012/12/19/storify-users-in-2012-posted-554000-stories-which-were-viewed-367-million-times/

Willnat, L., & Weaver, D. (2014). The American journalist in the digital age. Retrieved from http://news.indiana.edu/releases/iu/2014/05/2013-american-journalist-key-findings.pdf

Yakin, I., & Tinmaz, H. (2013). Using Twitter as an instructional tool: A case study in higher education. *The Turkish Online Journal of Educational Technology, 12*(4), 209–218. Retrieved from http://www.tojet.net/

ASSIGNMENT 10.1. LIVE TWEETING (CAMPUS LECTURE SERIES)

Reporters today no longer simply attend an event, take notes, and shoot either a picture or video for their stories. In the past, print reporters showed up with a notebook in hand and perhaps a camera, took notes, and then wrote their story on deadline. Broadcast reporters crowded near the stage with their cameras and equipment, got the video, and then reported the story live from the event or quickly produced a package in the live truck. But in today's world, both sets of reporters are not only taking notes and shooting pictures/videos, they also are "live-tweeting" the event to their audience.

The practice of live-tweeting will teach you how to take notes, not only on the lecture but also on how you can summarize the events happening in real time. You then will use these tweets to build a Storify that takes the best of the tweets from this class as well as anyone using the (official hashtag) and merge those with the narrative (your notes) about what the speaker told the crowd.

Assignment: Attend the Presidential Lecture. Be sure to get there early to get a good seat and to start talking to people who are gathered for the speech.

- Send out a tweet to your followers alerting them that you will be live-tweeting this event. Ask them to respond to your tweets or offer their own observations if they are attending the event.
- All tweets must use the official University Communications–approved hashtag.
- All tweets must be written in proper AP style and use correct grammar. No taking texting shortcuts to cut down words to fit the 140-character limit.
- No strange hashtags.
- Write what you see and what you hear. Write descriptions before the event of the excitement, the atmosphere, any unusual things in the audience, etc.
- Think of this as an exercise in taking notes. Tweet any quotes from the speech. Tweet your summary of key points. Tweet anything you observe in the crowd.
- Tweet directly through your phone, tablet, or laptop. It is easier to tweet from your computer, phone, or tablet than writing notes and then tweeting.
- Each tweet must be legitimate and related to the event. Don't tweet about your best friend or a party you went to over the weekend.
- You must post at least 15 legitimate tweets. Minimum: 15 tweets; unlimited maximum.
- Pictures: TwitPic.
- After the event, you will compile the tweets into a Storify. The Storify about the event is due at the beginning of class on Wednesday. We will link them on the class Facebook page.

ASSIGNMENT 10.2. THE TWIVINEAGRAM

The TwiVineAGram combined Twitter, Vine, and Instagram with Storify for a fun social and multimedia assignment. Journalists must know how to write stories for the legacy products such as print or broadcast as well as online and increasingly social and mobile media. Social media tools such as Twitter, Vine, Instagram, and Storify allowed reporters to use these tools to tell multimedia stories Access these tools on your smartphone or mobile device (tablet).

This year, we looked at several themes, and you can choose one that suits you the best. These questions should be a starting place for you to devise your own questions based on your individual story.

1. Black Friday recap

    a. What is the biggest must-have item?
    b. How early did people get to the stores?
    c. Did the stores in your area (name that area in the text somewhere) have any issues with people camping out, trampling anyone, etc.?
    d. Find two or three people and talk to them and shoot them with an interview on Instagram as they talk about the craziness of Black Friday.
    e. What was the atmosphere of the craziness of the day?
    f. Why in the world do people get up with the earliest of the early-rising chickens to do this?

2. Sophomore Study Skills Survival Guide to Finals Week

    a. Where is the best place to study on campus?
    b. What advice can your peers give about how to study for finals and survive?
    c. Where is the best place in the college town to get coffee or other caffeine for Finals Week Fuel?
    d. How can students balance the need to study with the need to sleep?
    e. How important is it to carve out time to take care of yourself during Finals Week?
    f. What are the best places to eat during that week? What type of food is the best for studying?

3. All About Basketball

    a. What makes a basketball game such an experience?
    b. Who are the key players this year?
    c. What's our outlook for conference play?
    d. Describe the atmosphere during game day.
    e. What's the best concession stand food?
    f. What's the best tradition at basketball games?
    g. Who are some of the biggest fans?

Assignment details: Choose one of the above topics. You should interview people for these questions as well as your own questions and then tweet their answers either as direct quotes or as videos (Vine, Instagram). You must have a minimum of 10 text tweets and five photo/video (Vine/Instagram) tweets. For the Vine and Instagram, you could try for mini-packages that show a variety of shot selections (wide, close-up, medium, establishing, etc.). The Instagrams and Vines may be shot on your smartphone or another mobile device. The tweets must carry the hashtag #JMC304 so that your peers (and the instructor) can find them.

You should embed the tweets and Vine/Instagram into Storify and then write context in the narrative spaces as if you are writing a news story. We examined Storify in class prior to this assignment. The Storify should read like a news or feature story, but it has to have the social media multimedia twist. You may choose the story structure in which to write your story (i.e., inverted pyramid, *Wall Street Journal* format, hourglass, a news feature, etc.). You also may use tweets and videos gathered by your classmates as long as they relate to your individual subject. You then must weave those tweets and visuals together with your own for a seamless story.

CHAPTER ELEVEN

Using Network Analytic Tools to Teach Social Media Impact on Citizen Journalism

K. HAZEL KWON

Journalism scholars have observed the rise of citizen journalism in the past two decades. Debates have revolved around how to define citizen journalism as a counterpart to the traditional form of professional journalism. The empowering potential of citizen journalism was particularly highlighted in the early and mid-2000s, during which a few landmark events lit up the power of online social networks in facilitating citizen engagements in news production and dissemination (Bruns, 2005; Chang, 2005). Discussions have ranged from attributing citizen journalism to a subversive collective power that could take on the mainstream news industry, to differentiating it from mainstream journalism as a hyper-local reportage or as a digital version of community or church newsletters, and to trivializing it as a collection of amateur musings (Jarvis, 2008; Lemann, 2006).

Along with the ubiquity of social media services in recent years, however, ordinary citizens' participation in creating and distributing news has become routine to the extent that the notion of "citizen journalism" no longer warranted a radical connotation. Since the rise of social media, the debate about the raison d'être of citizen journalism appears to have become somewhat outdated since ordinary user engagement in the news propagation process (such as re-creating, retelling, or re-disseminating), either deliberatively or unwittingly, has become much easier and plebeian compared to the previous online environment. Nonetheless, the prevalence of layperson involvement in the process of news propagation has not been translated into the frivolity of roles played by citizen journalism for social change. On the contrary, social media has allowed citizens to conveniently contribute in creating a new genre of news streams, often referred to as "affective news" (Hermida, 2010; Papacharissi & de Fatima Oliveira, 2012), into which ostensibly discrete types of information—such as factual news, emotional public utterances, pundit commentaries, and interpersonal casual conversations—have been blended and converged (Papacharissi & de Fatima Oliveira, 2012).

The role of citizen journalism has grown even greater and influential for political mobilization as evidenced by many contemporary social movements (Bennett & Segerberg, 2012; Castells, 2013). Likewise, citizen participation in social media information streaming has become critically important in collective sense making in times of crisis when mainstream or official sources of information might not perfectly function (Oh, Agrawal, & Rao, 2013). For example, moments after the Boston Marathon bombing incident in April 2013, the Boston police blogs crashed, cell towers were overloaded, and communications broke down. At that time, according to Cheryl Fiandaca, chief of the Boston Police social media department, Twitter became the communication lifeline between citizens and public officers, through which Boston Police could spread public safety information, answer requests for information, and react to misinformation in collaboration with citizen reporters (Fiandaca, personal communication, 2013). In this regard, journalism students in recent years have been playing a dual role, both as future journalism professionals and as informed citizens whose communication activities construct an important part of contemporary journalism ecology. Understanding the emergent media environment could help prepare students to carry out the dual roles more judiciously. This chapter introduces one of the methodological approaches in teaching the networked media environments, focusing in particular on a structural analysis of online social networks. Specifically, the chapter discusses how network analytics can be incorporated into the pedagogy of citizen journalism education and introduces a user-friendly, free computing program, NodeXL, as a useful tool for hands-on learning of social media networks in a classroom.

In sum, using a network analytic tool such as NodeXL helps journalism students learn information diffusion processes among ordinary online citizens particularly via social media platforms. Provided that information dissemination has been a nontrivial part of citizen journalistic activities as well as the information landscape in a networked individualism society, understanding online network structures that transmit news items is an insightful learning agenda for journalism students. Network properties such as centrality, density, and clusters can be easily explored by using social media data retrieved from NodeXL.

LITERATURE REVIEW

Citizen Journalists as Networked Individuals

A general consensus has been that citizen journalists might play a distinctive role from professionals in that they voluntarily participate in at least one of the stages of news content creating, processing, and distribution in forms of online activities (Bruns & Highfield, 2012). While early discussions have underscored the "alterity" imbued in the notion of citizen journalism, with the imagery of grassroots movements among rebels against the corporate-run media system, Goode (2009) pointed out that such heroic narratives hinder students from deeper understanding of the ways in which contemporary journalism has been shaped.

Goode (2009) stated that journalism "is essentially a craft of re-telling stories rather than simply disclosing them" (p. 1290). The retelling practice should not be divorced from the process of public sense-making and construction of meaning. In this sense, ordinary online users' practices of remixing and redisseminating existing mainstream media content must not be disregarded from the discourse on citizen journalism. Dvorak (2006) supported this claim by bringing out the notion of meta-journalism, which pertains to online user activities that virally mutated extant news contents including rating, commenting, tagging, sharing, or reposting. In a similar vein, gatekeeping literature has observed the emerging role of end users, or audiences, in the chain of information processing (Barzilai-Nahon, 2008; Kwon, Oh, Agrawal, & Rao, 2012; Shoemaker & Vos, 2009) and highlighted how networked audiences' sharing activities affect the information flow. For example, Kwon et al. (2012) demonstrated that news about international political conflict had spread among the online public via redistribution processes in social media platforms rather than via direct acquisition from mainstream news websites. Kwon et al. (2012) highlighted the potential that networked citizens could possess in reinforcing or disrupting news flows. More recently, Neuman, Guggenheim, Jang, and Bae (2014) used big data from various online news sources to illustrate that social media users' metajournalistic activities could reversely affect the agenda setting of issues in the mainstream media. Furthermore, scholars have emphasized the impact of users' networked sharing on journalism by referring to it as "audience gatekeeping" (Kwon et al., 2012; Shoemaker, Johnson, Seon, & Wang, 2011; Shoemaker & Vos, 2009) or "intermedia agenda setting" (Meraz, 2011; Sweetser, Golan, & Wanta, 2008).

Citizen journalism has not been necessarily rebellious against corporate influence. In contrast, much of these user engagements have leveraged profit-driven social media sites. Pew Internet Research (2012) indicated that the majority of U.S. adults were exposed to and engaged in disseminating political news via commercial social networking sites, which accounted for many top websites that were most trafficked throughout the Internet sphere (Top, 2014). Goode (2009) contended that user activities in these sites were a nontrivial part of citizen journalism practices, simultaneously benefited and constrained by network computing capabilities and profitability of social media corporations.

A large portion of recent citizen journalism activities has embraced metajournalistic practices such as commenting, sharing, liking, reposting, and so forth. The acts of metajournalism have mostly been occurring in popular, commercial social media platforms, for example, Facebook, Twitter, YouTube, Google Plus, and Instagram. These social media platforms have been the core enterprises in contemporary media environment. Here, user networking has been an essential common feature across different social media platforms. Without being connected to one another, citizens could not have shared perspectives on what their favorite news item was, why they liked it, what other news items were related to it, and how to re-create new content based on it. In other words, citizen journalists could only be prolific as networked individuals (Wellman et al., 2003). Social media have become an indispensable tool for citizen journalists by conveniently providing interactive opportunities with both already known social contacts as well as the broader public.

Social Media Network Analysis with NodeXL

According to Rainie and Wellman (2012), the proliferation of networked individuals (and networked individualism) was not solely caused by communication technology innovations. The steadfast societal changes toward greater social and physical mobility, globalization, and blurred community boundaries accounted for the ongoing shift from bureaucratic modernity into networked individualism. Nonetheless, the recent Internet and mobile revolution evidently accelerated this transition. Especially in the field of information and creative industry, the role of networked individuals has become even more prominent as their participation has collectively constructed new forms and channels for knowledge and information distribution. Citizen journalism could be one example of these products.

In this sense, understanding social network structures that underlie news dissemination could be a valuable learning objective for contemporary journalism education. Since many social media services have been operated by collaborations among computer programmers based on open sources or open codes to some degree, such openness enabled researchers and learners to access various social media data and better understand, for example, how users were interconnected with one another and how such interconnectedness facilitated the flow of information. While some students might have been equipped with programming language skills to utilize the Application Programming Interface (API) to get such data, the majority of journalism students might not be experts in computer programming. Hence, accessing such data could be a daunting task despite the acknowledgment of the necessity and the value of understanding the dynamics of social news dissemination.

To alleviate these challenges, certain software programs have become freely available in the recent years to provide non-experts with the tool for data access and analysis. This chapter focuses on one of these freely available software programs, NodeXL, which has been created and maintained by a collaborative academic research group since 2011 (Hansen, Shneiderman, & Smith, 2011). NodeXL (2014) is a free, open-source template for Microsoft Excel, which allows users to retrieve social media data, visualize the network of online users or news topics, and descriptively analyze characteristics of the network. NodeXL's interface is similar to conventional Excel templates.

The NodeXL website maintains an active network of scholars and learning community and also provides useful teaching resources. The usability of NodeXL is noteworthy for two reasons. First, it is based on a Microsoft Excel template, in which non-programmers can easily learn the interface designs and adopt functionalities for empirical explorations. Second, the program incorporates the plug-ins of several popular social media services, including Facebook, Twitter, LinkedIn, Flickr, into its platform, making it easy to access such social media data. Once these data are retrieved from social media platforms to NodeXL, students can readily conduct descriptive content and network analysis and visualization. NodeXL is intended to apply basic graph-theoretical network analysis to social media data. For a graph-theoretic understanding of the network, some basic knowledge regarding network constituents and measurements is essential. More details regarding the network constituents and measurement properties follow in the Applications section.

METHOD

A course titled Social Media Networks exemplified how the use of the network analytic tool could be incorporated into teaching a journalism undergraduate major. A network-based perspective played an increasingly important role in students' understanding of contemporary communication process. Especially, along with the popularity of social media, social networking has become a key term that characterizes online information consumption and dissemination.

The course has been offered every spring semester since 2013. It was a computer lab–integrated course and each student needed personal guidance to some extent. Therefore, the class was kept to a small or mid-size with no more than 40 students. The course accomplished three learning objectives. First, students learned the role of social technologies in shaping citizen journalism within a larger societal context. Second, students acquired hands-on skills to analyze and interpret the structure of social media news and user commentaries. Third, students enhanced their self-awareness as citizen journalists in digitally networked global communities. While the course was designed for an upper-level undergraduate course, it could be modified as a graduate-level course depending on the difficulty of the analytic problems the course intends to cover.

Keeping the networked media environment in mind, the course was designed into two parts. The first half of the semester aimed at theoretical learning: students were exposed to theoretical concerns and issues about the role of social technologies in evolving networked society, in particular how online networks were intertwined with media organizational changes and information consumption and creation processes. During this part of the semester, selected reading materials pertinent to social media–driven citizen journalism were assigned to students. Students were required to read the texts, attend lectures, and actively participate during in-class discussions. Students also performed home assignments including documenting a 24-hour log of their communication activities and media consumption and writing a personal essay after spending a day without using social media. The evaluation for this first part of the semester was based on two essay exams that covered lectures and readings, in addition to in-class activity participation and the quality of home assignments.

The second part of the course was methodological. During the second half of the semester, students learned social media network analysis and visualization (by using NodeXL) as part of citizen journalism training. The Applications section below mainly discusses this training. During the second part, students learned about the fundamental social network analytic properties and applied these properties to analyze social media data by using NodeXL. Students were trained intensively on how to use NodeXL to retrieve social media data from Facebook and Twitter, create and interpret network analytic results from the data, and visualize the network structure. During the training phase, each student's performance was reviewed on a weekly basis to make sure that students fully understood the training materials of the week. The evaluation of the second part was based on the weekly reviews and two assignments described in the Applications section below.

While the rest of this chapter mainly describes the analytic training, the theoretical discussion was as equally important as the hands-on training in promoting the understanding of the relationship between social media uses and the process of media content dissemination. During the first half of the semester, students had learned about networked individualism theory (Rainie & Wellman, 2012), audience gatekeeping theory (Barzilai-Nahon, 2008; Shoemaker & Vos, 2009), intermedia agenda setting theory (Meraz, 2011; Sweetser et al., 2008), and the concept of metajournalism (Goode, 2009). Through networked individualism theory, students learned how sociological changes have contributed to the advancement of communication technologies and mobile media. Students were also encouraged to view online information sharing practices not just as media consumption behaviors but also as a defining characteristic of being networked individuals. Theories of metajournalism, audience gatekeeping, and intermedia agenda setting were introduced to demonstrate the emergent roles of news audiences and the empowerment of citizen journalists in diversifying the journalism industry. After learning about such theories, the class reviewed recent case studies that exemplified social changes promoted by social media–driven citizen journalism such as the Arab Spring, the Occupy Wall Street movement, the Iceland Revolution, the Hong Kong Umbrella protest, and so forth.

The first part of the course—lectures on citizen journalism–related theories and case studies—could be flexible and versatile depending on the depth and width of knowledge of the instructor and the social currency of relevant issues. That said, training on network analytics were more technical and required the use of a predefined software program in learning processes. Therefore, demonstration of training outcomes in the next section was intended to provide insights for instructors interested in incorporating network analytic components into their courses.

APPLICATIONS

Social Network Analysis: Basic Concepts and Properties

Understanding basic network constituents and measurement concepts were prerequisite of leveraging social network perspective to train students on citizen journalism. In this course, students first learned two essential network constituents: nodes and edges. *Nodes* (interchangeably called *actors* or *vertices*) referred to the entities that constitute the network. While nodes commonly referred to individual persons, the unit of nodes could be defined in various ways including nonhuman entities depending on the research context. For example, nodes could represent media outlets, websites, and organizations. The more nodes included in the network, the larger the network size. Another element was *edges* (interchangeably termed as *links*, *ties*, or *relationships*), defined as the connections between a pair of nodes. Again, the substance of edges could be different depending on the topic of interest. It could represent, for example, friendship, communication frequency, node similarity, co-commenting, hyperlinking, and many others. For example, in Twitter, edges could represent who retweeted whom (retweeting relationships) or who mentioned whom (mentioning relationships).

To leverage the NodeXL program effectively, students needed to understand not only the nature of these network elements (i.e., what constitutes nodes and edges) but also the measurement properties widely used in the field of network analysis. In this course, the three most popularly used network properties were taught (committed learners were strongly recommended to study other available measures that help explore different aspects of network structures): "centrality," "density," and "clusters" (Wasserman & Faust, 1994). These three properties were chosen since they were the most widely used in understanding a variety of topics discussed among the journalism community, such as how to identify opinion leaders in the online personal networks and through which relational channels news stories could be spread in a social media platform.

Centrality. The centrality concepts "attempt to quantify the prominence of an individual actor embedded in a network" (Wasserman & Faust, 1994, p. 169). Students were educated on the centrality measures representation of how importantly an individual was positioned in a given network. Different types of centrality could be measured based on varied definitions of "important positions." First, an individual could be understood to be important if he or she had many direct edges with others (defined as degree centrality). Specifically, in-degree centrality refers to the incoming direct edges from the others' nominations; out-degree centrality refers to the outgoing direct edges from the focal actor to the others. Alternatively, a focal actor could be considered important if he or she had a direct connection to highly influential people who had a high degree centrality (defined as eigenvector centrality). For another definition of centrality, an individual could be considered important if he or she frequently lied on the paths between the other individuals (defined as betweenness centrality). According to Wasserman and Faust (1994), betweenness centrality "represents whether a particular actor might be able to control interactions between pairs of other actors in the network" (p. 188).

Different centrality measures could be used depending on situations. For example, if the news issues were hyperlocal, individuals who had many direct connections (degree centrality) in their local neighborhood might play an important role; at other times, people with high betweenness centrality might warrant attention especially if sensitive information were to be spread between two discrete social groups. In this case, people with high betweenness centrality would be likely to play as gatekeepers. In other cases when there was a need to mobilize resources for collective actions, identifying individuals with a high eigenvector centrality might be efficient since recruiting them would allow the campaigner to leverage more influenceability.

Density. Another important property could be the measure of density, referred to how densely members in a given network were interconnected. Computationally, it was the proportion of existing edges out of the maximum number of edges possible in a given network. The range of density value could be between zero and one, with the value of zero meaning no connections at all among the nodes, and the value of one meaning that every node is connected to every other node. Social media networks have shown much lower density values (close to zero) than interpersonal networks because the network size in social media would be generally much larger than in an interpersonal setting.

Examining network density would help understand different network structures across various social platforms. For example, suppose that Facebook network showed a higher density compared to Twitter, given the same network size. The result would suggest a different social mechanism underlying each platform. Facebook might be similar to in-person neighborhood dynamics where personal social relationships and acquaintances would be intertwined, whereas Twitter might be akin to a broadcasting media system, in which issue attention would matter more than social relational maintenance so that networked users could be strangers to one another (for instance, a follower of a media network personality on a Twitter account would have no personal relationship with such a person even though the two were connected on Twitter).

Cluster. Lastly, students learned the concept of cluster, which referred to a subgroup within a network. Multiple subgroups could be identified in a single network. The clustering results would show a systematic pattern in the act of connecting, representing a prominent rule for group formations. The rule might be rooted in political orientation, geographical proximity, educational level, occupational type, religious background, and others. For example, in Blogosphere, politically liberal bloggers would tend to keep a blogroll of likeminded liberal blogs, whereas the conservative bloggers would exhibit the opposite list (Hindman, 2008). Looking at clustering in a social media network could help determine whether information was exchangeable across different clusters of people.

Centralities, density, and cluster are automatically computable by using network analysis software such as NodeXL. After learning about social network constituents (i.e., nodes and edges) and basic property measurements (i.e., centralities, density, and cluster), students practiced analyzing online social media data retrievable by using NodeXL. As of 2014, retrievable social media data from NodeXL were Facebook (Fan Page, Group, and Personal Network), Twitter (List, Search, Users Network), Flickr (Users and Tags Network), and YouTube (Users and Video Network). Data accessibility might vary contingent upon changes in each of the social media companies' privacy policies.

Two of the in-class assignments that students had engaged in were presented for the use of NodeXL in a classroom. Each assignment was a two-week project. The first assignment aimed to understand how to identify potentially important "opinion leaders" based on network positional properties. The majority of students in this particular course have shown interest in exploring their Facebook personal network structure as an example of online social network. Therefore, the following demonstration was based on the Facebook personal network analysis. The second assignment aimed to explore Twitter messages related to an important news issue, and understand the scope and impact of citizen journalism in the process of news propagation. Twitter was a proper venue for this course considering its improvised news reporting system (Hermida, 2010) as well as its potential to harness community intelligence during a social crisis (Oh et al., 2013). Therefore, Twitter was a proper social media outlet for students to learn about the role of citizen reporting.

Facebook Personal Network Analysis

Once familiarized with the interface and the basic functions of NodeXL, students carried out the first assignment, which requested that they analyze their own personal network on Facebook (see Assignment 11.1). One prerequisite was that students must have an account on Facebook. The procedure was as follows: First, students logged into the Facebook system by using the plug-ins embedded in the NodeXL, which allowed them to import their own personal network data. Network data from Facebook included the list of friends, connections among, and personal attributes (such as gender, hometown, relationship status) of the friends when publicly available in Facebook. Second, students used the social network analysis function to compute the measures of basic properties. The results were saved as additional attributes alongside personal information. Third, students visualized their Facebook personal network by adjusting the network layouts, colors, size, and opacity. Visualization was customizable for the sake of the students' best interpretation of their network structures. Lastly, students reported in detail the structural characteristics of their Facebook personal network in terms of centralities, density, and clustering, and discussed who should be considered important actors (or opinion leaders) in the process of information spread within the network. Students often found it surprising that some social connections existed among individuals who were affiliated with discrete social groups, and that these crosscutting edges tended to result in high betweenness centralities. Based on the results of various degrees of centralities and sub-clusters, students could understand the interconnections among users and interpret which structural positions manifested the important roles in spreading the information.

Basically, this assignment intended to allow students to learn about their online personal network structure in one of the social media platforms (i.e., Facebook) and identify the locations of individuals who could be the potential opinion leaders based on the analysis of group clustering and node centralities. Considering that online news contents were largely carried over through personal social media networks, learning about the structure of media content flows on an interpersonal level could be an insightful, self-reflective project for journalism students.

Citizen Journalism Practices in Twitter During the Egyptian Revolution in 2011

For the second assignment, students carried out the analysis of Twitter data (see Assignment 11.2). NodeXL allowed students to use their Twitter account to search and download publicly available tweets (and retweets) as well as user networks based on retweeting or mentioning relationships. Students could collect the data by inputting search keywords of interest into the Twitter import bar embedded in the interface of NodeXL program. When inputting the search keywords, students set the maximum number of tweets to download (allowable up to 20,000 tweets). According to Twitter company policy, the Streaming API-based data collection had to be in "real time," which could span from the day of data collection up to the past two weeks. Twitter import function in NodeXL was programmed based on the Streaming API, adhering to Twitter data policy. Therefore, students were instructed to choose one of the most recent trending news topics and used the related search keywords for data collection. The downloaded dataset included tweets (and retweets) that contained the search keywords, the time of tweets sent, user IDs and profile descriptions, and the mentioning and/or retweeting relationships among users. For network analysis, users were defined as nodes, and mentioning and/or retweeting relationships were defined as edges.

After experimenting with Twitter data that the students had collected for their own interest, students performed the second assignment, which requested that they analyze the data provided by the instructor. This data was the collection of tweets during the first three days of the Egyptian Revolution, January 24–26, 2011. The Egyptian Revolution in 2011 demonstrated a successful role of citizen journalists who used social media, including Twitter, in disseminating of information and mobilization (Castells, 2013). It was an exemplary news topic for exploration. Whereas the real-time Twitter data collection in the NodeXL was versatile for recent news agendas, data from the Egyptian Revolution in 2011 was historical and not accessible by the real-time collection in NodeXL. Therefore, students were given the dataset previously collected by the instructor. Because NodeXL had not yet been released as of January 2011, the instructor had scripted Python (a computer language) to gain access to Twitter Streaming API. Although the instructor had used Python, the data collection process was almost identical to that of NodeXL considering that both Python and NodeXL were based on Streaming API. Specifically, the dataset was collected in real time and contained the same types of information such as tweets and retweets, time of tweets being sent, user IDs, user profile description, and retweeting relationships.

The three-day dataset of Egyptian Revolution 2011 included about 4,000 tweets: Among them, the majority was in English, and some were in Arabic. A small portion of tweets was in other languages. Only English and Arabic tweets, which were translated into English by a bilingual graduate student, were further investigated. The data were stored chronologically in a real-time flow. Students worked in teams (two or three) to explore the nature of tweet messages, characteristics of the message senders, and geo-locations from which the tweets were generated. Upon completion of the review, students conducted a network analysis to determine the most central users in spreading the news to the Twitter public about the incident. Based on the exploration of data, students produced the following results.

Types of Communication. In terms of message types, the most prominent practice was to provide mass media coverage or official sources that affirm the importance and severity of the demonstrations, followed by the real-time situational updates, tactical knowledge, and expressive support for the protesters. Interpersonal use of Twitter was the least prominent despite the fact that the majority of tweets were sent out by personal users who were not affiliated with any mainstream media organization. Students explored user profiles and concluded that more than 70% of the identified tweet senders were personal users—citizen journalists. Only about 12% were media or governmental organizational representatives, and just a few tweets were from computer-bots. About 10% of the tweets were sent from unidentified users.

Out of the total tweets, 39% (n = 1,728) included URL, which hyperlinked a resource from the broader websphere. Among the hyperlinked resources, the majority were coming from user-generated content websites as much as from professional media websites, again confirming that the citizen journalists played a prominent role along with professional media sectors. While some were hyperlinked from non-movement or non-advocacy related organizations, there were not many resources directly distributed from the movement, advocacy, or campaign organizers.

Density. The retweet network analysis showed that the social network grew from 269 nodes with 405 edges to 900 nodes with 1,692 edges. Due to a relatively large size (hundreds of users participating in tweeting about the revolution), the density score tended to be very low, implying the sparseness of the network. Despite the sparseness, however, most of the actors were connected in some ways and together formed a big, single cluster. These results implied that Egyptian Revolution activists were organically networked through active information exchanges via retweeting practices.

Centrality. Students also identified the most central users by using centrality measures (i.e., degree centrality), concluding that prominent actors were personalized Twitter users who did not claim any association with advocacy/movement/media organizations. Only eight of the users among the top 25 out-degree centralities were identified as institutional Twitter users representing either media organizations or movement-related organizations. Also, the highest in-degree centralities were associated with citizen journalists or user-generated resources from social media sites such as YouTube, Storify, TwitPic, Facebook, and yfrog. While reputational media websites such as BBC and Guardian and their representative Twitter accounts were also paid a non-trivial portion of attention, there were only six actors in the list that represent movement/advocacy/campaign organizations.

In sum, this assignment aimed to allow students to explore the real-world citizen journalistic practices occurring in social media platforms. By interpreting and analyzing tweet contents, user profiles, and retweet networks, students could examine the role of ordinary online citizens in producing and spreading information during an important social, political event.

DISCUSSION

Ordinary citizens' influence on the process of content production, gatekeeping, and agenda setting has increased along with the popularity of social media services (Kwon et al., 2012; Neuman et al., 2014). The process of creating, editing, re-creating, and distributing user-generated content has become so convenient and commonplace in this networked public environment that the subversive connotation that used to characterize citizen journalism a decade ago appeared somewhat out of fashion. Nevertheless, ordinary users' everyday information sharing practices in social media has contributed to result in a new genre of information production, so called affective news (Papacharissi & de Fatima Oliveira, 2012), which has combined facts with emotional reactions, and professional commentaries with laypeople opinions. Affective news has become an important resource for the online public to collectively understand an uncertain situation, often providing the impetus for political mobilization and crisis management.

In this chapter, citizen journalists were characterized as networked individuals who were embedded in online social network structures. Social network analysis was demonstrated in the course titled Social Media Networks as a way of training students in contemporary citizen journalism. Students learned that network structural properties shaped by various online networking activities such as friending, following, retweeting, sharing, liking, co-commenting, and so forth were as important as message contents in understanding, identifying, and interpreting the nature of citizen journalism. In particular, students utilized one of the user-friendly social media network analysis programs, NodeXL, for training. Among various social network analytic tools currently available to the academic community, NodeXL was selected due to (1) its availability at no cost, (2) its easy-to-learn interface on Excel template basis, and (3) its ability to allow users to retrieve some of the most popular social media data.

While the two assignments exemplified in this chapter were designed for a 400-level journalism undergraduate major course, they could also be adapted for the graduate level. Basic concepts of social network analysis, including nodes, edges, centrality, density, and cluster, were introduced. Based on these concepts and measures, the goal of the first assignment was to analyze a personal Facebook network to understand the structural properties of potential opinion leaders in spreading news in an online network setting. The goal of the second assignment was to explore how ordinary users participated in the process of news streaming during a real-world news event. For the second assignment, students engaged in teamwork to analyze Twitter message contents, user characteristics, and the structure of retweet networks during the Egyptian Revolution in 2011.

As the course assignments exemplified, learning social network perspectives and acquiring hands-on skill on network analysis software such as NodeXL could help journalism students explore not only the online users' everyday communication practices (such as the use of Facebook personal network) but also online news propagation during an extreme event (such as the use of Twitter hashtags and retweets). Students learned that the contemporary social media environment allowed online users to play a role in citizen journalism, either consciously or unwittingly, by engaging in retelling and redisseminating processes. Social media users could often play the role of audience gatekeepers, whose activities might collectively affect the salience of news agenda. Indeed, Neuman et al. (2014) highlighted that social media users' attention to news issues prominently affected the coverage of some issues, despite not all, in traditional media outlets. The "reverse patterns" of agenda setting (Neuman et al., 2014, p. 204) from social media users to mainstream journalism has become more prevalent. In this context, learning about the online network dynamics under which diverse types of information have been ceaselessly transmitted could help contribute to journalism students' understanding of the milieu in which information tools and social networks have been integrated.

REFERENCES

Barzilai-Nahon, K. (2008). Toward a theory of network gatekeeping: A framework for exploring information control. *The Journal of the American Society of Information Science and Technology, 59*(9), 1493–1512.

Bennett, W. L., & Segerberg, A. (2012). The logic of connective action: Digital media and the personalization of contentious politics. *Information, Communication & Society, 15*(5), 739–768.

Bruns, A. (2005). *Gatewatching: Collaborative online news production.* New York, NY: Peter Lang.

Bruns, A., & Highfield, T. (2012). Blogs, Twitter, and breaking news: The produsage of citizen journalism. In R. A. Lind (Ed.), *Produsing theory in a digital world: The intersection of audiences and production in contemporary theory* (pp. 15–32). New York, NY: Peter Lang.

Castells, M. (2013). *Networks of outrage and hope: Social movements in the Internet age*. Malden, MA: Polity Press.

Chang, W. Y. (2005). Online civic participation, and political empowerment: Online media and public opinion formation in Korea. *Media, Culture & Society, 27*, 925–935.

Dvorak, J. C. (2006). Citizen journalism is like citizen professional baseball: You can't play pro baseball just because you think the Seattle Mariners stink. *PC Magazine, 25*(18): 1.

Goode, L. (2009). Social news, citizen journalism, and democracy. *New Media and Society, 11*(8), 1287–1305.

Hansen, D., Shneiderman, B., & Smith, M. A. (2011). *Analyzing social media networks with NodeXL: Insights from a connected world*. Chicago, IL: Morgan Kaufmann.

Hermida, A. (2010). Twittering the news: The emergence of ambient journalism. *Journalism Practice, 4*(3), 297–308.

Hindman, M. (2008). *The myth of digital democracy*. Princeton, NJ: Princeton University Press.

Jarvis, J. (2008). In Munbai, wittnesses are writing the news. *The Guardian*. Retrieved from http://www.theguardian.com/media/2008/dec/01/mumbai-terror-digital-media

Kwon, K. H., Oh, O., Agrawal, M., & Rao, H. R. (2012). Audience gatekeeping in the Twitter service: An investigation of tweets about the 2009 Gaza conflict. *AIS Transactions on Human-Computer Interaction, 4*(4), 212–229.

Lemann, N. (2006). Amateur hour: Journalism without journalists. *The New Yorker*. Retrieved from http://www.newyorker.com/archive/2006/08/07/060807fa_fact1

Meraz, S. (2011). Using time series analysis to measure intermedia agenda-setting influence in traditional media and political blog networks. *Journalism & Mass Communication Quarterly, 88*(1), 176–194.

Neuman, R. W., Guggenheim, L., Jang, S. M., & Bae, S. Y. (2014). The dynamics of public attention: Agenda-setting theory meets big data. *Journal of Communication, 64*(2), 193–214.

NodeXL. (2014). Retrieved from http://nodexl.codeplex.com

Oh, O., Agrawal, M., & Rao, H. R. (2013). Community intelligence and social media services: A rumor theoretic analysis of tweets during social crises. *MIS Quarterly, 37*(2), 407–426.

Papacharissi, Z., & de Fatima Oliveira, M. (2012). Affective news and networked publics: The rhythms of news storytelling on# Egypt. *Journal of Communication, 62*(2), 266–282.

Pew Research Center's Internet & American Life Project. (2012, March 12). *Social networking sites and politics*. Retrieved from http://pewinternet.org/Reports/2012/Social-networking-and-politics.aspx

Rainie, L., & Wellman, B. (2012). *Networked: The new social operating system*. Cambridge, MA: MIT Press.

Shoemaker, P. J., Johnson, P. R., Seon, H., & Wang, X. (2011). Readers as gatekeepers of online news: Brazil, China, and the United States. *Brazilian Journalism Research, 6*(1), 55–77.

Shoemaker, P. J., & Vos, T. P. (2009). *Gatekeeping theory*. New York, NY: Routledge.

Sweetser, K. D., Golan, G. J., & Wanta, W. (2008). Intermedia agenda setting in television, advertising, and blogs during the 2004 election. *Mass Communication & Society, 11*(2), 197–216.

Top 500 sites on the web. (2014). Retrieved from http://www.alexa.com/topsites

Wasserman, S., & Faust, K. (1994). *Social network analysis: Methods and applications*. New York, NY: Cambridge University Press.

Wellman, B., Quan-Haase, A., Boase, J., Chen, W., Hampton, K., Díaz, I., & Miyata, K. (2003). The social affordances of the Internet for networked individualism. *Journal of Computer-Mediated Communication, 8*(3). Retrieved from http://onlinelibrary.wiley.com/doi/10.1111/j.1083-6101.2003.tb00216.x/full

ASSIGNMENT 11.1. FACEBOOK PERSONAL NETWORK

Students were asked to analyze positional properties of individuals embedded in a Facebook personal network and identify who could be potentially important "opinion leaders" based on the network analysis results.

Specific Instructions Were as Follows:

- Step 1. Download Facebook personal network by using "import" function embedded in the NodeXL interface. Nodes were the student's friends, and edges were friending relationships among the friends.
- Step 2. Compute analytic measures (centrality scores, density, and clustering) by using the network analytic functions embedded in NodeXL interface.
- Step 3. Compare and interpret the results by sorting the nodes list based on the ranks of each computed measurement property. Students could easily compare the network positional properties of each node. Some nodes were highly ranked according to one metric (e.g., betweenness centrality), while others were more highly ranked according to another metric (e.g., degree centrality). Students were asked to make a separate note about these results for their own record so that they could keep track of the results.
- Step 4. Visualize the network graph by adjusting the visual properties of nodes (i.e., size, color, opacity, and shape) and edges (i.e., width, color, and shape). If the network seemed to be too large or too cluttered, students were recommended to use the "dynamic filters function" embedded in NodeXL interface, by which less important nodes could be removed from the graph image.
- Step 5. Write a reflection paper that contains the network graph image (and the description about the process of visualization) as well as the network analysis results and the process of decision making in identifying opinion leaders. In particular, students were asked to describe (1) the structure of their Facebook network (e.g., size, density, number of clusters, social background or context of clustering), (2) network analytic rationales on which they concluded that some actors were the more important than others, and (3) the roles of these important individuals in spreading information by exemplifying a fictional situation wherein certain information needs to be disseminated.

ASSIGNMENT 11.2. INTERPRETING CITIZEN JOURNALISM PRACTICES IN TWITTER DURING THE EGYPTIAN REVOLUTION IN 2011

The second assignment aimed for students to demonstrate their understanding of the social network analytic concepts/measurements, and their capability to apply them to the real-world Twitter data. The assignment was composed of three parts:

(1) Network visualization (creating an effective graph image),
(2) An image caption, and
(3) Interpretation/discussion of Twitter data analysis results.

As a final assignment of the semester, this assignment asked students to creatively use the functions and tools embedded in Node XL and carefully study tweets and user profiles to interpret a Twitter network in the most meaningful way.

While the step-by-step instructions were identical to Step 2 to 5 of Assignment 11.1, the requirements for the reflection paper of the second assignment had some variations. Specifically, the reflection paper must (1) describe the network structure, (2) provide network analytic reasoning on identifying important actors, and (3) discuss the role of an ordinary citizen reporting in disseminating news about the Egyptian Revolution in 2011 by offering specific examples drawn from the analysis of the dataset.

CHAPTER TWELVE

Going ON A News Consumption Diet

Engaging Students in Meaningful Current Events Discussions Through Social Media

MELONY SHEMBERGER

At a time when social media and other digital technologies are prevailing in the information world, the need for news literacy is growing. In fact, mobile news consumers especially are more likely to follow news frequently, use multiple news platforms/sources, seek practical utilities from news information, value portal news sites, share content with others, and be receptive to advertising campaigns, especially those via social media (Vagg, 2010). Meanwhile, social networking has increased in popularity, as the percentage of Americans using Facebook has increased from 8% in 2008 to 51% in 2011 (Arbitron, 2011). Yet, despite a growing realization that media literacy is important in the modern age of media convergence, the development of media literacy education at the university level has been slow (Schmidt, 2013).

Given this current landscape, students—many of whom have mobile devices and access social media—from all academic disciplines should know what is happening in current events. However, regardless of all the digital transformations in recent years, today's students do not seem interested in current events and are not engaged with traditional news sources. About three-quarters of college students, typically ages 18–24, consume news through digital platforms such as news websites and social media rather than more traditional sources (Center for the People & the Press, 2012). These students are part of the "always on" world of interactive media and have different expectations and learning styles than previous generations (Baird & Fisher, 2005). Because students rely heavily on the Internet for information, they might not identify what they see as news in the traditional sense. Mihailidis and Cohen (2013) noted that young adults "refer to social networks, aggregators, and mobile apps for all information habits, instead of siloing out specific media for news, politics, personal communication, and leisure" (p. 5).

However, the ability to find, identify, and use news content is part of media literacy—and having such proficiency is an industry standard. One thing is clear: Knowledge of current events is critical to embrace media literacy; therefore, instructors need an "always on" pedagogy that relies on more relevant methods. The purpose of this chapter is to provide ideas on how to structure current news assignments using a digital, interactive approach. The discussion and the assignments in this chapter are centered on the application of the engagement pyramid social media model, which serves as the guiding framework to explore how social media tools are used creatively by both the instructor and the student for rich discussions of current events. Details for assignments are also offered for educators who might be searching for innovative ways to boost media literacy and current news discussions in their classrooms.

LITERATURE REVIEW

Media Literacy

Before pedagogical techniques to increase a student's knowledge of current events can be discussed, the concept of media literacy needs to be examined. With the advent of new media technologies, the landscape of media has changed considerably (Lin, Li, Deng, & Lee, 2013), so much so that educators have had to redesign their pedagogy for classroom use and student learning effectiveness. College students today have been part of the culture of digital natives. However, while access to social media technology and the Internet has had a positive impact on all levels, including personal, family, professional, and social, such media have not affected any observed increase in media literacy (Garcia-Ruiz, Ramirez-Garcia, & Rodriguez-Rosell, 2014). The development of media skills among citizens, especially young people and children, to exercise a critical and active role in relation to the media is a key development in this society of "media prosumers" (Garcia-Ruiz et al., 2014). This means if digital technology enabled members of an audience to exert influence, then media literacy should be a priority for effective decision-making and citizenry.

This raises the need to discuss what media literacy is and discover some of the issues associated with it. Media literacy has been defined in a number of ways. With myriad applications, there were several ways to understand and define media literacy (Fedorov, 2003). Potter (2010) identified at least 23 definitions that have been developed by scholars and citizen action groups. Generally, the definitions have suggested that media literacy mixes core competencies with a variety of peripheral skills (Schmidt, 2013).

Martens (2010) dated the concept to the mid-1980s. Media literacy may be applied as a tool to foster and promote active citizenship. Most initiatives have taken place in K–12 education (Hobbs, 2004; Martens, 2010). At colleges and universities, media literacy education is considered important by faculty, but few are involved in teaching the subject (Schmidt, 2012). In fact, formalized, comprehensive media literacy education programs in postsecondary higher education have been uncommon (Mihailidis, 2006; Silverblatt, Baker, Tyner, & Stuhlman, 2002; Wulff, 1997). Departments of mass communication, journalism, media studies, and education (Mihailidis, 2008) often had to weave the material into the curriculum. News consumption in college has been important for students to practice because these years could mold lifelong habits (Diddi & LaRose, 2006) when it came to voting and other important decisions.

Media literacy has appealed to students at different education levels. Given this variance, this chapter defines the concept as "the ability to access, analyse, evaluate and create messages across a variety of contexts" (Livingstone, 2004, p. 5). The term is related to news literacy. Scholars might debate some slight differences between the concepts. In fact, Koltay (2011) said that it could be difficult to think of media literacy as different from other types of literacy, such as information literacy, digital literacy, and emerging technology literacy. In some literature, *news literacy* and *media literacy* are used interchangeably. Throughout this chapter, the two terms are treated as the same because both concepts are concerned with evaluating and using information responsibly in all contexts of journalism and digital technology.

Regardless of the term associated, for media literacy to grow, students must become news consumers, not "news grazers" who check news on the Internet or through social media feeds for the latest updates but do not read the news article to digest the entire story (Diddi & LaRose, 2006). Compared to habitual news consumers, the Pew Research Center described grazers as "considerably younger, less interested in serious news, and use media sources at lower rates except for cable and online news. Moreover, fewer grazers than habituals say they enjoy keeping up with the news" (Center for the People & the Press, 2002, p. 4). By 2010, almost 74% of adults ages 18–29 said they stayed on top of the news in this grazing manner (Center for the People & the Press, 2012). News grazers have learned about the news through networking sites such as Facebook, but most of these users noticed news when on the site for other reasons (Matsa & Mitchell, 2014). However, news grazers often followed current events in one or more of these five categories: entertainment (73%); people and events in the community (65%); sports (57%); national government and politics (55%); and crime (51%) (Matsa & Mitchell, 2014).

Students also have been "power searching" their way through the news in the same manner that they were power-searching their way online—never searching deeply, purposefully, or effectively (Cheney, 2010). That's because students have wanted news that is accessible and convenient (Jarvis, Stroud, & Gilliland, 2009). Their news mostly came from the Internet and comedy television, but students also have been able to differentiate among other types of media sources, such as cable news, daily newspaper, news radio, local TV news, network news, and news magazines (Jarvis et al., 2009). Jarvis et al. (2009) concluded that "college students may have a more layered and unmeasured appetites for news than has previously been acknowledged" (p. 37).

Although college and high school students might use a variety of different media to access news, they have placed news consumption at a lower priority than other daily needs (Huang, 2009). Consequently, students preferred short stories and multimedia content because they were immediately satiated. Furthermore, these students "like to control what news to access and how to access it. For some, they do not want to be passive news receivers; they also want to contribute content via participatory media" (Huang, 2009, p. 117). This explanation has been consistent with findings in other studies. Peters (2012) contended that to understand how audiences engage with journalism in the contemporary age, scholars should conceive of news consumption not just as something people do, but as something done in a particular mobile place or platform, such as Facebook or Twitter. It considered the experiences of consuming journalism, and reflected on the influence "space" exerted in this equation.

Nevertheless, these findings also have implied that as college students were exposed to more information online, their knowledge of current events became more fragmented. Before the Internet and World Wide Web altered the news industry, journalism instructors often used current events quizzes to require their students to find and consume news through traditional channels such as print newspapers and local and network television news programs. Now, students face "the integration of various information types (news, entertainment, personal communication) and mediums (television, radio, print) into aggregated spaces. Search engines and social networks have replaced specific channels, shows, and even websites as the predominant places youth go for information" (Mihailidis & Cohen, 2013, p. 2). All users could customize the web to their own tastes through the sites they chose to visit, follow, "like," or subscribe.

Basically, instructors have adopted college students' news preferences for social media and the web as opportunities to encourage them to increase their current news consumption. They also have placed students into the role of content creator. Rheingold (2008) considered social media a teaching tool, writing that "participatory media literacy is necessarily a hands-on enterprise, requiring active use of digital media by students" (p. 25). Other studies have reported a similar approach, including Herrington and Parker (2013), Lee and McLoughlin (2007), Mihailildis and Cohen (2013), and Livingstone (2004).

METHOD

Engagement Pyramid Social Media Model

Quizzes that test a student's knowledge of current news and events have been popular among journalism instructors. Success on these quizzes and related assignments required students to read newspaper articles, watch TV newscasts, or listen to radio newscasts—all of which can be found online in some format. This is where the application of social media in assignments and exercises can help encourage a healthy news consumption diet for today's college student. Such assignments and activities also help students to identify and minimize information overload. Holton and Chyi (2012) found that the majority of today's news consumers felt overloaded with the amount of news they encountered, largely in part because of news organizations' increased volume of production and delivery platforms. To understand how social media can be an instructional tool for media literacy, the engagement pyramid social media model can serve as the theoretical framework in this discussion and as the basis for designing pedagogy. The assignments and exercises, ideal for any undergraduate mass communication or journalism course, explained in this chapter were designed using the engagement pyramid social media model and were used in two separate undergraduate journalism courses at a university in the Southeast. The courses were basic news writing and copy editing. The model along with the details of the assignments and exercises are explained in the next section.

The engagement pyramid model has been used in leadership, marketing, education, and organizational settings. Although the model did not apply directly to student engagement, it could serve as the underlying principle when journalism educators structure assignments that would lead to enhanced media literacy skills among students. Li (2010) developed the theory to offer a simple view of the different ways social media are used to engage with an audience. The model followed a five-step process: watching, sharing, discussing, creating, and curating. Each step is explained below, and implications are offered for educators.

Watching. People who watch their social media accounts do so to absorb content, such as blogs, videos, or status updates. They then use the information to make decisions, learn from others, or be entertained. To apply this to an educational setting, instructors should understand the content that this group has been consuming and create content that would engage watching.

Sharing. People who post status updates on social sites such as Facebook and Twitter also share photos, videos, and articles. They do this to offer information to their peers, both to support others and to demonstrate knowledge. To engage with students, educators should simplify and innovate sharing possibilities with suitable social media tools.

Commenting. Audience participants respond to others' content by posting comments, reviews, or ratings. They also support or contribute ideas and opinions. Educators should make certain that the selected social media tools used would allow for commenting. Comments should foster an open and friendly environment.

Creating. People who create and publish their own content do so because they want to express their identities, be known for their material, want to be heard, or desire recognition. To engage, educators should have a platform where students could have their voices heard. Educators also could provide public recognition to students who proved to be the most helpful in the online learning community.

Curating. Those who curate are involved in online communities because they invested time and communication in the success of a product, service, or community and want to be recognized for their work. From an educational perspective, instructors should consider students as curators in online assignments and other relative classwork and recognize them publicly for their contributions.

Each of Li's social media model's levels contributed to the overall media creation process, which involved planning, producing, and sharing original media (Ascher & Pincus, 1984; Hobbs, 2004; Lund, 1998; O'Brien, 2005; Williams & Medoff, 1997). Media creation competencies have been important for a number of reasons. One would be the necessity to participate in the modern world of media convergence, in which individuals have been part of an active culture involving engaged, participative consumers who would create, share, and seek out new information (Jenkins, 2006). Others have suggested that the development of media creation competencies has been important because individuals could grasp the conventions of professional media by gaining experience with creating their own media (Hobbs, 1998; Sefton-Green, 1999).

APPLICATIONS

Applying the Engagement Pyramid Model to Assignments and Exercises

As much as journalism educators would favor the addition of news literacy as a required course for their programs, it would not always be feasible to create a semester-long news or media literacy course. Giving students the critical-thinking skills necessary to discern what would be credible information in the dominant social and digital world would be the mission of news literacy; however, single courses have not always been the answer. Instructors could help students achieve news literacy by incorporating fun and interactive assignments and classroom exercises on a frequent basis to foster competencies and attitudes among students, most of whom have grown up surrounded by the digital world. In many ways, the instructor should design lessons around the core idea that students have been an "end user" who would be conducting most of their learning and career development outside the traditional classroom (Baird & Fisher, 2005). For the journalism student, the assignments described in this chapter would play a significant role in positioning the student toward a healthy news consumption diet that would facilitate that learning and development.

In spring 2014, students in a basic news writing course and an intermediate copy editing course in an undergraduate program at one university in the Southeast followed current events closely with the integration of social media techniques in the classroom. To demonstrate how social media could be used to encourage a student's news consumption diet through the engagement pyramid social media model, this case study explored three practical activities that educators could use. There have been many social media tools, with many more under development that could be integrated into the curriculum; the ones discussed in this chapter were among a few. Educators are encouraged to find those social media and digital tools with which they have been most familiar. Each exercise included details and recapped briefly how the engagement model connected students and media literacy.

Daily Current Events Discussion

Every day, students in both news writing and copy editing courses had to come to class with current events items to discuss. Students appeared to be more attracted to crime news and the unusual, but even such news could be scrutinized. After a brief discussion, the instructor then shared a digital digest of the top news students also should track in case new developments occur by the next class meeting.

The instructor used the social media curation tool Storify to create a daily digest of topics in the news to begin each class session with a brief, yet robust, discussion of current events. Storify helped make sense of what people would post on social media. Users curated the most important voices and turned them into stories. Topics discussed in the instructor's Storify digest ranged from major crime news to international affairs to media ethics. Before class each day, the instructor reviewed news feeds from major news organizations' social media accounts and websites to determine major news of the day that students needed to know and likely did not catch their attention. At the start of class, the instructor first asked the students what they knew to be happening in the news. Top student responses included news items relating to crime, entertainment, and unusual or odd happenings. Students also commented on whether the news stories demonstrated effectiveness (such as quality of sources and story structure) and how the message contained in the tweet or Facebook post was presented (such as clarity of message).

After a few minutes, the instructor then showed the Storify string of current events on the projection screen to engage in further conversation. Students also were able to see any accompanying photos or videos with the items. In addition, students were encouraged to visit the instructor's Storify site to review items in more detail outside of class time because questions about current events appeared on quizzes and tests.

Engagement Pyramid. The educator served as the key influencer in this activity. With the five steps—watching, sharing, commenting, creating, and curating—the instructor invited students also to engage in meaningful current events discussions. The instructor gathered current news content that students should know and curated content that engaged watching. Students became part of the process when they watched for news updates, shared their insights, and commented on their peers' contributions.

Following Current Events in Copy Editing

News consumption today is an active experience—for journalists, readers, and viewers. Curated online reporting is efficient and interactive, and it might be a responsibility for copy desk editors, whose responsibilities might include creating and curating digital material for a story online or for additional content for the reader or viewer. At the start of class for several weeks in the semester, copy editing students led discussions on expanded topics, such as national issues, global issues, legal matters, education, business, sports, music, weather, and military. Each student selected a topic from a list generated by the instructor and selected a presentation date. In a class with 15 students, the following topics were presented:

> Legal, Sports, Business/economy, Religion, Books/literature, Science/technology, Health/fitness, International issues, National government, State/local government, Military, Employment/unemployment, Music, Education, and Weather

These topics were broad in scope, but students were given a number of options to pursue. For instance, if a student chose international issues as the topic, Storify could focus on coverage of an isolated event, such as the United Nations General Assembly speeches and session. Likewise, a student who signed up to research current events in education might focus only on higher education. Another approach could be to study the top issues covered in the news in education; poverty, testing and assessment, and career readiness might be the major themes to explore. The sky was the limit with this assignment.

This assignment required the students to pull a variety of credible online content—such as tweets, YouTube videos, and Facebook posts—into a digital digest for presentation. Students were assessed in four categories with an equal number of points assigned: logical arrangement of content, balance of narrative and online content, variety of sources, and presentation. Because this was the main project for the copy editing class, the Storify assignment was awarded a total of 200 points.

Since the focus was on engagement with current news, students had to choose news items that were no more than two weeks old. However, the best way to tackle the assignment was to begin about a week before the presentation. This way, a student had the most current information to share. Students could start watching, creating, and curating at any time; they would simply have to edit their posts frequently, but there was nothing wrong with that.

This time frame allowed flexibility for students whose topics might not yield daily news. In addition, inclusion of content from reputable news organizations—the Associated Press, the *Washington* Post, and *USA Today*, among others—was mandatory. Storify was used as the curation method. This exercise enabled the students to act as digital content editors, reflective of the multimedia duties copy editors now oversaw on the copy desk.

Along with sharing and discussing current news during these presentations, the use of these three social media tools allowed opportunities to discuss objective reporting, grammar, accuracy, news source credibility, search engine optimization (SEO) headlines, and other topics relevant to journalism skills. See Assignment 12.1 regarding the overview and instructions for this assignment.

Engagement Pyramid. All five levels were followed. Students watched social media accounts belonging to news organizations for content. Students created a solid presentation about their topic using timely content and shared the news with the class, allowing for discussion to flow. Students were encouraged to access the Storify posts curated by their peers as reference tools for quizzes and examinations.

Weekly Current Events Assignment in News Writing

Each week in the news writing course, students followed and reported current news using a digital tool called Fotobabble to record their scripted news story over a related photo. Fotobabble is a free cloud-based social media app that allowed students to use a photograph secured from an online news site and then narrate their own story. The tool has achieved prominence in the public relations industry, and some news organizations have used the tool on their websites as sidebar content to a story or as a standalone feature similar to a still photo. The instructor also suggested news outlets for students to find and read news. The students then worked on their own time to find articles and write and record their stories in Fotobabble. In the final step, they embedded the HTML code from Fotobabble to link their work on a course blog powered by Blogger.

Specifically, students had to find a news story that had suitable visuals, to re-write the story into their own words, and then to record the 60- to 90-second story over a photograph uploaded to Fotobabble. The resulting "story" was embedded into the course blog in Blogger. Students also had to write a headline and to give a credit line to the news source. The headline consisted of more than three or four keywords to grab a reader's attention, just as a headline would in a newspaper or online publication.

A 50-point rubric was created to assess the students' work in five areas: newsworthiness relevance of topic, photo selection, headline and photo credit, presentation, and content. Students also were required to listen to their peers' work, which gave students greater insight into current events and created a class news product on which to base discussions. Students were encouraged to comment on story structure, flow, or newsworthiness. The deadline fell on Tuesday to complete the Fotobabble and Thursday for the comments.

During Thursday's class, the instructor selected five of the week's top stories and opened the class session with an interactive poll, using PollEverywhere.com. Students texted their vote from their cell phones, and the live poll results were displayed on the projection screen. After the results were determined, the class discussed why the winning story merited designation as the most newsworthy story of the week. Because the Fotobabble assignment was more complex than the Storify activities, the step-by-step instructions are included in Assignment 12.2.

Engagement Pyramid. This exercise empowered the students throughout each step. Most students watched their social media accounts for content, making the Fotobabble application effective. Students created and shared the current event they chose for the week and also commented on other posts submitted by their peers. The course blog acted as a curation tool to house the students' work.

Student Insights

Near the end of the semester, students were surveyed via an anonymous online questionnaire that asked them to provide comments, thoughts, and suggestions about the various current events assignments and exercises.

Starting each class with brief current events discussions using Storify showed success. Students favored the discussions, and their responses illustrated the effectiveness of the daily activity: "It gave me insight into certain news articles I may not have followed otherwise," a student wrote. The Storify approach also enabled students to see news produced in various ways and formats.

The copy editing students also found their Storify current events projects purposeful. Moreover, some students grasped the significance of the project. "Thanks to the technology that we have today, news is available at your fingertips 24/7 and that is something every student should take advantage of," one student wrote.

Most students in the news writing course said they enjoyed the Fotobabble assignment. One student wrote in the survey, "I liked that everyone had a different story to report and it made me more aware of the world around me. And it was fun to be a reporter of some sort."

These assignments were not without limitations. During both semesters, several students had issues with browser compatibility (i.e., Chrome, Firefox). A few news writing students completed their assignments on their smartphones using the Fotobabble and Blogger apps, but most students chose to use a laptop computer with a built-in microphone. To make students aware of technology needs, the instructors have specified in the syllabus that students need to have access to a computer with a built-in webcam and microphone as well as to either a tablet or smartphone with texting capabilities. Storify also experienced minor maintenance issues at times for the instructor, but the copy editing students did not experience any technical problems while using the digital curation tool.

DISCUSSION

Media literacy competencies are necessary for full participation in the modern world, and such competencies are increasingly necessary for today's journalism students, who must be able to engage, interact, and connect with news sites to read, consume, and produce news content. Because 41% of young adults ages 18–24 are more likely to read news through digital and online news sites (Center for the People & the Press, 2012), college educators should try include digital and social media learning tools as part of the media literacy discussion in their classrooms. Thirty-four percent of that age group said news is consumed through social media sites (Center for the People & the Press, 2012), and that statistic could grow in the years to come.

Social networking media provide the opportunity to take engagement to deeper levels as well as to address learning styles rooted in digital technologies (Baird & Fisher, 2005). Tandoc (2014), in a study based on a survey of 364 undergraduate journalism students, looked at how news consumption patterns influence the journalistic role conceptions that students hold. The study found that students rated the interpreter role as most important. Students who prioritized the interpreter role also tended to get their news from online sources and social media (Tandoc, 2014).

However, since students use the Internet for social media and news consumption, it would be possible that they have become "news grazers" (Diddi & LaRose, 2006). Students may read a tweet from a news organization, but they need to know there is more to it than that. They have to click the link and read the story—not only to see what the news story is reporting, but also to examine journalistic writing critically and learn how to write in the manner professional journalists do. Students are already engaged with social media, but by requiring them to write their own stories and upload them through social media channels, they are practicing the levels outlined in the engagement pyramid social media theory.

Furthermore, students are engaged with a network at various levels, as mentioned by Li (2010). The Fotobabble assignment, for instance, helped students connect not only to major news sources but also to each other; they discussed current events, learned how to produce brief broadcast news stories, and used HTML coding to embed those stories into the course blog. These tasks helped them to connect not only to the larger news community but also to their peers as they completed the weekly assignments and provided feedback to each other.

An important point about assessment needs to be stressed here. While the use of social media tools can greatly provide students with opportunities for all five levels of the engagement pyramid model, it is imperative that the use of these tools be tied closely to learning objectives. Although the instructor used Fotobabble for one of the assignments, there are several platforms and other digital exercises that journalism educators could consider and integrate into their curriculum. The key is to make certain the tool selected reflects the purpose or objective of the assignment or lesson. In other words, educators should be careful not to use social networking tools for the sake of using social technology, and should keep in mind how the use of any type of technology element can support student learning. Regardless of the tool chosen, as the engagement pyramid model demonstrated, a learner engaged in creating and re-creating his or her own learning network built deeper levels of understanding. In this case, the students began to understand the categories of news and question why a particular story (breaking news, etc.) might be deemed more newsworthy than a story drawn from another category.

Finally, the students who embraced the assignments not only increased their understanding of media literacy, they also saw their quiz grades improve as they engaged better with the news material in fun and interactive exercises. While academic achievement is certainly important, student enthusiasm in tracking current news is just as crucial in the development of a successful journalist or any other communication professional. Journalism educators, however, should shape their pedagogical use of digital tools in such a way that the practice of staying on top of current news is expected of students, so that consuming a balanced news diet becomes the norm.

REFERENCES

Arbitron. (2011). The infinite dial 2011: Navigating digital platforms. Retrieved from http://www.arbitron.com/downloads/infinite_dial_2011_presentation.pdf

Ascher, S., & Pincus, E. (1984). *The filmmaker's handbook: A comprehensive guide for the digital age.* New York, NY: Penguin.

Baird, D. E., & Fisher, M. (2005). Neomillenial user experience design strategies: Utilizing social networking media to support "always on" learning styles. *Journal of Educational Technology Systems, 34*(1), 5–32.

Center for the People & the Press. (2002). *Public's news habits little changed by September 11.* Retrieved from http://www.people-press.org/2002/06/09/publics-news-habits-little-changed-by-september-11/

Center for the People & the Press. (2012). *In changing news landscape, even television is vulnerable: Trends in news consumption 1991–2012.* Retrieved from http://www.people-press.org/2012/09/27/in-changing-news-landscape-even-television-is-vulnerable/

Cheney, D. (2010). Fuzzy logic: Why students need news and information literacy skills. *Youth Media Reporter, 4,* 45–47.

Diddi, A., & LaRose, R. (2006). Getting hooked on news: Uses and gratifications and the formation of news habits among college students in an Internet environment. *Journal of Broadcasting & Electronic Media, 50*(2), 193–210.

Fedorov, A. (2003). Media education and media literacy: Experts' opinions. In *Mentor. A media education curriculum for teachers in the Mediterranean.* Paris, FR: UNESCO. CD-ROM (English Version).

Garcia-Ruiz, R., Ramirez-Garcia, A., & Rodriguez-Rosell, M. M. (2014). Media literacy education for a new prosumer citizenship. *Comunicar, 22*(43), 15–23. doi:10.3916/C43-2014-01

Herrington, J., & Parker, J. (2013). Emerging technologies as cognitive tools for authentic learning. *British Journal of Educational Technology, 44*(4), 607–615.

Hobbs, R. (1998). The seven great debates in the media literacy movement. *Journal of Communication, 48*(1), 6–32.

Hobbs, R. (2004). A review of school-based initiatives in media literacy education. *American Behavioral Scientist, 48*(1), 42–59.

Holton, A. E., & Chyi, H. (2012). News and the overloaded consumer: Factors influencing information overload among news consumers. *Cyberpsychology, Behavior & Social Networking, 15*(11), 619–624. doi:10.1089/cyber.2011.0610

Huang, E. (2009). The causes of youths' low news consumption and strategies for making youths happy news consumers. *Convergence: The International Journal of Research into New Media Technologies, 15*(1), 105–122.

Jarvis, S. E., Stroud, N. J., & Gilliland, A. A. (2009). College students, news use, and trust. *Communication Research Reports, 26*(1), 30–39.

Jenkins, H. (2006). *Convergence culture: Where old and new media collide.* New York, NY: New York University Press.

Koltay, T. (2011). The media and the literacies: Media literacy, information literacy, digital literacy. *Media, Culture & Society, 33*(2), 211–221.

Lee, M. J. W., & McLoughlin, C. (2007). Teaching and learning in the Web 2.0 era: Empowering students through learner-generated content. *International Journal of Instructional Technology and Distance Learning, 4*(10), 21–34.

Li, C. (2010). *Open leadership: How social technology can transform the way you lead.* San Francisco, CA: Jossey-Bass.

Lin, T.-B., Li, J.-Y., Deng, F., & Lee, L. (2013). Understanding new media literacy: An explorative theoretical framework. *Educational Technology & Society, 16*(4), 160–170.

Livingstone, S. (2004). Media literacy and the challenge of new information and communication technologies. *The Communication Review,* 7(1), 3–14.

Lund, D. J. (1998). Video production in the English language arts classroom. *The English Journal, 87*(1), 78–82.

Martens, H. (2010). Evaluating media literacy education: Concepts, theories and future directions. *Journal of Media Literacy Education, 2*(1), 1–22.

Matsa, K. E., & Mitchell, A. (2014). *8 key takeaways about social media and news.* Retrieved from http://www.journalism.org/2014/03/26/8-key-takeaways-about-social-media-and-news/

Mihailidis, P. (2006). Media literacy in journalism/mass communication education: Can the United States learn from Sweden? *Journalism and Mass Communication Educator, 60*(4), 416–428.

Mihailidis, P. (2008). Are we speaking the same language? Assessing the state of media literacy in U.S. higher education, *Simile 8*(4), 1–14.

Mihailidis, P., & Cohen, J. N. (2013). Exploring curation as a core competency in digital and media literacy education. *Journal of Interactive Media in Education, 2.*

O'Brien, A. (2005). Filmmaking across the curriculum. *Screen Education, 38,* 87–90.

Peters, C. (2012). Journalism to go. *Journalism Studies, 13*(5/6), 695–705. doi:10.1080/1461670X.2012.662405

Potter, W. J. (2010). The state of media literacy. *Journal of Broadcasting & Electronic Media, 54*(4), 675–696.

Rheingold, H. (2008). Using social media to teach social media. *New England Journal of Higher Education, 23*(1), 25–26.

Schmidt, H. C. (2012). Essential but problematic: Faculty perceptions of media literacy education at the university level. *Qualitative Research Reports in Communication, 13*(1), 10–20.

Schmidt, H. C. (2013). Addressing media literacy within higher education: A comparison of faculty and student perceptions. *Northwest Journal of Communication, 41*(1), 133–159.

Sefton-Green, J. (Ed.). (1999). *Young people, creativity and new technologies: The challenge of digital arts.* London, UK: Routledge.

Silverblatt, A., Baker, F., Tyner, K., & Stuhlman, L. (2002). Media literacy in U.S. institutions of higher education. Retrieved from http://www.webster.edu/medialiteracy/survey/survey_Report.htm

Tandoc, E. C. (2014). The roles of the game: The influence of news consumption patterns on the role conceptions of journalism students. *Journalism & Mass Communication Educator, 69*(3), 256–270. doi:10.1177/1077695813520314

Vagg, T. (2010). The mobile news audience will be highly receptive to ad campaigns. *New Media Age,* August 19, 2010. Retrieved from http://www.nma.co.uk/opinion/industry-opinion/analystspeak-the-mobile-news-audience-will-be-highly-receptive-to-ad-campaigns/3017246

Williams, S. H., & Medoff, N. J. (1997). Production. In W. G. Christ (Ed.), *Media education assessment handbook* (pp. 235–254). Hillsdale, NJ: Erlbaum.

Wulff, S. (1997). Media literacy. In W. G. Christ (Ed.), *Media education assessment handbook* (pp. 123–142). Hillsdale, NJ: Erlbaum.

ASSIGNMENT 12.1. STORIFY CURRENT EVENTS PROJECT

News consumption has morphed into an active experience. Curated online reporting is efficient and interactive—and it might be a component of the copy desk.

For this assignment, you will use the topic you choose to create a Storify post. Each student will lead the current events discussion at the start of class. The first one begins in two weeks.

These topics are broad in scope, but you have a number of options to pursue. For instance, if you have international issues as the topic, you can choose to focus on the events involving the ISIS. You also can choose a number of news angles and organize them in a certain way. The sky is the limit.

The best way to tackle this assignment is to begin about a week before you present. This way, you have the most current information to share. Of course, you may start any time; you will have to edit your post frequently, but there is nothing wrong with that. Try to use the most current information possible.

This assignment is worth 200 points, and your grade will be based on the following criteria:

- Logical arrangement of content (50 points)
- Balance of narrative and online content (50 points)
- Variety of sources (50 points)
- Presentation (50 points)

Create your Storify account, if you do not have one. If you have any questions, we can address them at the next class session.

ASSIGNMENT 12.2. DIRECTIONS FOR CREATING A FOTOBABBLE AND POSTING IT TO BLOGGER

Instructors: These directions can be included in a handout given to students. It also is recommended that some class time, about 15 to 20 minutes, be set aside for the instructor to walk through the directions with students in a computer lab or have students bring their laptops to class, if possible.

Directions for creating a Fotobabble and posting it to Blogger

1. Find a photo from a news site (CNN, NPR, network news, local online newspaper, etc.)
2. Take notes on what you want to say about the story. You are writing your own news story. This will allow you to have a more polished presentation.
3. Save photo to your desktop (for PC, right click; for Apple, drag onto desktop). NOTE: You will need to copy the URL address where you got the photo. This will go into your blog post.
4. Go to Fotobabble site and create an account to log into it. Check email to confirm your account.
5. Click on "Create New" at top.
6. Use the first option to obtain the photo from your computer.
7. Give your Fotobabble post a title that describes its content. Be sure to acknowledge the source of your photo in the title.
    a. Example: Protests taking place after the federal government shut down (Source: CNN.com)
    b. Change your setting to private
8. Record your Fotobabble. (Again, it will help to keyboard or write your "script" on paper first so that you don't get tongue-tied or ad lib.) Record for up to 60 seconds. To hear your recording, push Play. To start again, push Record and it erases everything you had to that point and starts over for you.
9. Save your Fotobabble.
10. In a separate tab or window, log in to the class blog. (Instructors: Blogger, WordPress, or a learning management system are effective platforms. Whichever one you choose, provide the course blog address here and send your students invitations to become "authors" on the blog.)
11. Begin a new post, using the "Compose" toggle button. Type the headline and source of your Fotobabble and your name.
12. Go to your Fotobabble page and go to the right side of the page and copy the HTML code. Do not use the iFrame embed code.
13. Go back to the blog post and toggle to the HTML code button. Paste the embed code from Fotobabble. It will look something like this:

    Example:

```
<object width="600" height="650"> <param name="movie" value="http://
www.fotobabble.com/mediafiles/templates/basicslideshow/fb.swf">
</param><param name="bgcolor" value="#ffffff"></param><param
name="wmode" value="transparent"></param><param name="allowFullScreen"
```

value=”true”></param><param name=”allowscriptaccess” value=”always”></param><param name=”FlashVars” value=”autoPlay=false&remoteXML=true&appURL=http://www.fotobabble.com&id=SCtqeHVORjltVzQ9&increment=true&layout=default&size=large”/><PARAM NAME=BASE VALUE=”http://0ztevmk3kfy73v4y4m82-fb-production-content.s3.amazonaws.com/content/SCtqeHVORjltVzQ9”/><embed src=”http://www.fotobabble.com/mediafiles/templates/basicslideshow/fb.swf” type=”application/x-shockwave-flash” FlashVars=”autoPlay=false&remoteXML=true&appURL=http://www.fotobabble.com&id=SCtqeHVORjltVzQ9&increment=true&layout=default&size=large” BASE=”http://0ztevmk3kfy73v4y4m82-fb-production-content.s3.amazonaws.com/content/SCtqeHVORjltVzQ9” allowscriptaccess=”always” bgcolor=”#ffffff” allowfullscreen=”true” width=”600” height=”650”></embed></object>

DO NOT ALTER THE CODING! This is an extremely important point. One change to the code, and your work will not display properly.

14. Click on preview to make sure the post works. When you are satisfied with the post, click Publish in the top right corner.
15. Submissions will be due every Tuesday, unless otherwise noted. You then will look at the blog and comment on up to three of your peers' entries until Thursday, when we will vote in class for the brief of the week. Comments should be constructive about the story. (Example: The story seemed to jump from one point to another, etc.) You must make three total comments each week.
16. In class Thursday, be ready to discuss the stories of the week and to vote. The top five stories of the week will be chosen by the instructor for the Thursday voting. The top winner will receive extra credit added to his or her lowest score. Bring your smartphone for text voting on PollEverywhere.com, if you wish to participate.

About THE Editor

Hana S. Noor Al-Deen (Ph.D., State University of New York, Buffalo) is a professor of communication at the University of North Carolina Wilmington. She has postdoctoral education in digital multimedia and 3D computer animation from three universities. Noor Al-Deen has been involved with digital technology for more than three decades starting with her doctoral dissertation, "Corporate Teleconferencing: Audio, Video, and Computer." She has been teaching courses in social media, digital multimedia, 3D animation, advertising, mass communication, and intercultural/international communication.

Noor Al-Deen was the lead co-editor of the books *Social Media and Strategic Communications* (2013) published by Palgrave Macmillan, United Kingdom and *Social Media: Usage and Impact* (2011) published by Lexington Books as well as the editor of the book *Cross-Cultural Communication and Aging in the United States* (1997) published by Lawrence Erlbaum Associates/Routledge. Also, her work has appeared in books published by Sage Publications, Greenwood Publishing, and Roxbury Publishing as well as in journals such as *International Journal of Advertising, World Communication Journal,* and *Global Media Journal.* Among her journal articles was "The Evolution of Rai Music" published by *Journal of Black Studies* in 2005 and republished by Sage Publications in 2011 in a "four volume distillation of the best that has been thought and published in the academic study of Popular Music." Noor Al-Deen was a consulting editor of the *International and Intercultural Communication Annual.* She was a book proposal reviewer for Sage Publications and Allyn & Bacon Publishing, a journal reviewer of *International & Intercultural mmunication*, a convention paper reviewer for the Association for Education in Journalism and Mass Communication (AEJMC), Newspaper Division as well as for various communication divisions at the Southern States Communication Association (SSCA). In addition to the United States, Noor Al-Deen has conducted research and presented academic papers in Canada, China, Egypt, France, and Turkey.

Finally, Noor Al-Deen has devoted more than 30 years of her life toward helping abused, homeless, and shelter animals. She is a proud and active member of various animal rights organizations such as the American Society for the Prevention of Cruelty to Animals (ASPCA), The Humane Society of the United States (HSUS), and Animal Legal Defense Fund (ALDF).

About THE Contributors

Thibault Adda (B.A., Paris West University Nanterre La Défense) is an international graduate student at the University of Central Florida's Nicholson School of Communication. Adda's research interests include crisis communication, social media and new technologies for corporate communication, Internet marketing, SEO, and SEM.

David Baines (M.A., Leicester University, UK) is a senior lecturer in journalism at Newcastle University, UK. He teaches multimedia journalism, media law and ethics, and editing and design. His primary research interests concern the transformation of local and hyperlocal journalism and the changing roles of journalism in communities.

Stanley J. Baran (Ph.D., University of Massachusetts) is professor and chair of the Communication Department at Bryant University. He taught for four years at Cleveland State University, eventually moving to the University of Texas. He led the Department of Radio-TV-Film's graduate program for six of his nine years in Austin and won numerous teaching awards there, including the AMOCO Teaching Excellence Award as the best instructor on that 40,000-student campus, the College of Communication's Teaching Excellence Award as that college's outstanding professor, and Utmost Magazine's Student Poll for best instructor. Dr. Baran moved to San Jose State University in 1987 and served nine years as Chair of the Department of Television, Radio, Film & Theatre. At SJSU he was named President's Scholar as the university's outstanding researcher. Among his other awards are his designation as among "the top 3% of active researchers in communication" in a 1993 Communication Education study of the field's journals, the Broadcast Preceptor Award for Excellence in the Literature of Mass Communication from San Francisco State University, an Emmy nomination for a dramatic television series he produced on mentally retarded people, and a Fulbright Fellowship spent at the Institut fur Journalismus und Kommunication at the Hochschule fur Musik und Theater in Hannover, Germany. Dr. Baran has published 10 books, scores of scholarly articles, and sits or has sat on the editorial boards of six journals. His work has been translated into half a dozen languages.

Rowena L. Briones (Ph.D., University of Maryland) is an assistant professor of public relations at the Richard T. Robertson School of Media & Culture at Virginia Commonwealth University. She teaches a number of different courses focused on public relations writing, research, campaign development, and social media, and her research interests explore the intersection of public relations, new media, and health communication.

Melissa D. Dodd (Ph.D., University of Miami) is Assistant Professor of Advertising-Public Relations at the University of Central Florida's Nicholson School of Communication. Her research focuses on social capital theoretical approaches; intangibles; corporate social responsibility/advocacy; and social media as related to the measurement of public relations.

Julia Daisy Fraustino (M.A., Pennsylvania State University) is a doctoral fellow in communication at the University of Maryland, College Park. In August 2015, she joined the Reed College of Media at West Virginia University as an assistant professor of strategic communication. She teaches several courses in strategic integrated communication, public communication campaigns, crisis/risk communication, and social media strategy, and her research investigates intersections of public relations, risk/crisis communication, ethics, and social media.

Holly Kathleen Hall (J.D., University of Arkansas at Little Rock) is an associate professor of strategic communication at Arkansas State University. Her research interests are in the areas of U.S./U.K. comparative media law and social media. She currently teaches classes in media law, social media, and public relations and is Accredited in Public Relations (APR) by the Public Relations Society of America.

Gwyneth Howell (Ph.D., University of Western Australia) is a professor of digital and social communication at the University of Western Sydney, Australia. Professor Howell's research focuses on crisis and disaster management and response techniques, social and digital communication, public policy, and covers a variety of topics (including natural disasters, corporate social responsibility, education, and urban animal management). She has published in journals including the *Journal of Business Research*, the *Journal of Global Academy of Marketing Science*, *Journal of Emergency Management*, and *Public Relations Review*.

Melissa L. Janoske (Ph.D., University of Maryland) is an assistant professor of journalism at the University of Memphis in Tennessee. She teaches a variety of public relations courses and crisis communication, and her research interests focus on the intersection of public relations, community building, and social media.

K. Hazel Kwon (Ph.D., State University of New York, Buffalo) is an assistant professor of communication studies at Arizona State University. Her teaching and research interests focus on social technologies, with a particular emphasis on the dynamics in which technology-mediated communication influences citizen engagement and public opinions. Her research was supported by NSF and HASTAC, and her recent publications appear in multiple journals including *Journal of Computer-Mediated Communication*, *Cyberpsychology, Behavior and Social Networking*, and *American Behavioral Scientists*.

Abbey Blake Levenshus (Ph.D., University of Maryland) is an Assistant Professor in the University of Tennessee's School of Advertising and Public Relations. She teaches and researches about organizational public relations, specializing in government PR, risk and crisis communication, and social media. Her professional experience includes serving as communication director for a U.S. congressman and companies in Washington, D.C.

Regina Lewis (Ph.D., The University of North Carolina, Chapel Hill), is a professor of communication at The University of Alabama. Her teaching and research interests focus on new media and consumer insights. Specific publications include "Understanding Pass-Along Emails: Motivations and Behaviors of Viral Consumers" and "Building Marketing Relationships on Twitter: A Content Analysis of University Twitter Accounts."

Rohan Miller (Ph.D., University of Western Australia) is a professor of marketing communication, digital marketing, and consumer behavior at the University of Sydney, Australia. Dr. Miller's research focuses on consumption with negative effects (e.g., gambling, alcohol, food, digital social games), marketing communication and digital communication, public policy, and it covers a variety of topics (including education, urban animal management, wool value chains, ethics, asbestos and the environment). He has published in journals including the *Journal of Business Ethics*, the *Journal of Business Research*, the *Journal of Marketing Communications*, *Marketing Theory*, and *Public Relations Review*.

Mary Jackson Pitts (Ph.D., University of Southern Mississippi) is a professor of media at Arkansas State University. Her research focuses on new media applications with current interest in iPad technology. She teaches graduate-level research and theory classes. Most recently she has worked with gifted and talented teachers promoting the use of iTechnology in the classroom with a specific emphasis on storytelling.

Melony Shemberger (Ed.D., Tennessee State University, Nashville) is an assistant professor of journalism and mass communications at Murray State University in Murray, Kentucky. Her research interests include pedagogy in journalism courses, historical explorations of notable journalists, and freedom of information issues. Her publications have appeared in various journals and she is a member of Phi Kappa Phi.

Kathleen Stansberry (Ph.D., University of Oregon) is a professor of communication at Cleveland State University. Her teaching interests include public relations and social media and her research addresses the intersection of public relations, online communities, and health.

Beth L. Sundstrom (Ph.D., University of Maryland; M.P.H., Brown University) is an Assistant Professor of Strategic Communication at the College of Charleston. She researches and teaches strategic health communication, social marketing, social media, and women's health. She has professional experience in public relations and continues to consult in strategic health communication.

Melissa Wall (Ph.D., University of Washington) is a professor of journalism at California State University–Northridge where she teaches and researches about citizen and social media. She is the editor of *Citizen Journalism: Valuable, Useless or Dangerous.*

Brandi Watkins (Ph.D., The University of Alabama) is an assistant professor of public relations at Virginia Tech. She teaches classes in public relations and social media. Her research interest includes studying the strategic use of social media for relationship building in the context of sports.

Leigh Landini Wright (M.F.A., Murray State University) is an assistant professor of journalism and mass communications at Murray State University. A former newspaper reporter and editor, her teaching and research interests involve best practices of social media in the classroom and in community journalism.

Index